VOLUME ONE HUNDRED AND THIRTY

PROGRESS IN MOLECULAR BIOLOGY AND TRANSLATIONAL SCIENCE

Molecular Basis of Olfaction

VOLUME ONE HUNDRED AND THIRTY

PROGRESS IN MOLECULAR BIOLOGY AND TRANSLATIONAL SCIENCE

Molecular Basis of Olfaction

Edited by

RICHARD GLATZ

D'Estrees Entomology & Science Services
MacGillivray, SA 5223, Australia
The University of Adelaide, Urrbrae, SA, 5064, Australia
South Australian Museum, Adelaide, SA, 5000, Australia

ELSEVIER

AMSTERDAM • BOSTON • HEIDELBERG • LONDON
NEW YORK • OXFORD • PARIS • SAN DIEGO
SAN FRANCISCO • SINGAPORE • SYDNEY • TOKYO
Academic Press is an imprint of Elsevier

Academic Press is an imprint of Elsevier
225 Wyman Street, Waltham, MA 02451, USA
525 B Street, Suite 1800, San Diego, CA 92101-4495, USA
125 London Wall, London, EC2Y 5AS, UK
The Boulevard, Langford Lane, Kidlington, Oxford OX5 1GB, UK

First edition 2015

ISBN: 978-0-12-802912-1
ISSN: 1877-1173

CONTENTS

CONTRIBUTORS

Jonathan D. Bohbot
United States Department of Agriculture, Beltsville Agricultural Research Center, Invasive Insect Biocontrol and Behavior Laboratory, Beltsville, Maryland, USA

Arthur de Fouchier
Institute of Ecology & Environmental Sciences of Paris, INRA, Versailles, France

Joseph C. Dickens
United States Department of Agriculture, Beltsville Agricultural Research Center, Invasive Insect Biocontrol and Behavior Laboratory, Beltsville, Maryland, USA

Jean-François Gibrat
INRA UR1077 Mathématique Informatique et Génome, Domaine de Vilvert, Jouy-en-Josas, France

Emmanuelle Jacquin-Joly
Institute of Ecology & Environmental Sciences of Paris, INRA, Versailles, France

Nicolas Montagné
Institute of Ecology & Environmental Sciences of Paris, UPMC-Sorbonne Université, Paris, France

Richard D. Newcomb
School of Biological Sciences, University of Auckland, and The New Zealand Institute for Plant & Food Research Limited, Auckland, New Zealand

Edith Pajot-Augy
INRA UR 1197 NeuroBiologie de l'Olfaction, Domaine de Vilvert, Jouy-en-Josas, France

Marie-Annick Persuy
INRA UR 1197 NeuroBiologie de l'Olfaction, Domaine de Vilvert, Jouy-en-Josas, France

Guenhaël Sanz
INRA UR 1197 NeuroBiologie de l'Olfaction, Domaine de Vilvert, Jouy-en-Josas, France

Jackson T. Sparks
United States Department of Agriculture, Beltsville Agricultural Research Center, Invasive Insect Biocontrol and Behavior Laboratory, Beltsville, Maryland, USA

Thierry Thomas-Danguin
INRA UMR 1129 Flaveur, Vision et Comportement du Consommateur, Dijon, France

Anne Tromelin
INRA UMR 1129 Flaveur, Vision et Comportement du Consommateur, Dijon, France

William B. Walker
Chemical Ecology Research Group, Department of Plant Protection Biology, Swedish University of Agricultural Sciences, Alnarp, Sweden

Guirong Wang
State Key Laboratory for Biology of Plant Diseases and Insect Pests, Institute of Plant Protection, Chinese Academy of Agricultural Sciences, Beijing, PR China

Dieter Wicher
Max Planck Institute for Chemical Ecology, Department Evolutionary Neuroethology, Jena, Germany

Jin Zhang
State Key Laboratory for Biology of Plant Diseases and Insect Pests, Institute of Plant Protection, Chinese Academy of Agricultural Sciences, Beijing, PR China

PREFACE

Smell is a potent wizard that transports you across thousands of miles and all the years you have lived.

Helen Keller

This poignant quotation by Helen Keller speaks to the evocative nature of olfaction for humans. Beyond being simply an important diagnostic mechanism for interpreting the environment, olfaction can often recall old memories or stir complex emotions. In my home country of Australia, there are stories of soldiers returning from battle in World War II by ship and realizing that they were nearing their homeland prior to sighting it, simply from the characteristic smell of the oil-laden *Eucalyptus* trees that dominate much of the Australian landscape. These weary combatants were not just detecting trees but imbibing their loved ones, their childhoods, their hopes, and their loss.

Coming from Helen Keller, this quote also subtly hints at the key role olfaction plays when sight is not the primary sense used for navigation. This is actually the case for most of the animals on earth; huge numbers of species of invertebrates use olfaction as their key method of assessing their environment and detecting food, mates, hosts, predators, etc. In creatures such as insects, olfaction-related cognition is much simpler than for humans; however, it is known to be important in individual learning, in parasitic wasps for example. Olfaction is so important to insects that they have evolved extremely sensitive olfactory receptors (ORs) to detect low concentrations (sometimes nanomolar and below) of volatile compounds; these receptors largely reside in their antennae but do occur elsewhere. The olfactory sensitivity of insects helps make them formidable evolutionary competitors but is also exploited by humans to disrupt insect behavior (e.g., pheromone disruption of moth pests and pheromone trapping).

Olfaction has attracted significant scientific interest for many years. In 1937, Japanese researchers utilized electrodes to measure the negative electrical potential generated across olfactory epithelium of dogs, caused by olfactory stimulation. This technique was adapted for study of frogs and rabbits in 1956 and given the name electro-olfactography; it has since been widely utilized for study of olfaction in mammals. In 1957, the technique was adapted to insects and named electroantennography, and in 1959 the first insect pheromones were characterized from the silk moth, *Bombyx mori*.

While electrophysiological techniques such as these were used successfully for decades and could be used to detect the presence and degree of olfactory stimulation by various compounds, they were unable to decipher the molecular basis of olfaction.

However, around the same time in 1953, Watson and Crick published the structure of DNA. This was a seminal moment in science and was built on by others to produce great advances in our understanding of molecular biology and in the power of the techniques available to study it. Then in 1991, Richard Axel and Linda Buck discovered that vertebrate ORs were a subclass of the well-known G protein-coupled receptor (GPCR) family of proteins. This discovery (which was subsequently recognized with a Nobel Prize in 2004) combined with advances in DNA/RNA sequencing technologies and bioinformatics led to the elucidation of OR repertoires of a range of vertebrate species and of associated molecular signaling processes. The first vertebrate receptor to be deorphaned (have its cognate ligands characterized) was OR17 from the rat in 1998, which was shown to react to C_7–C_{10} saturated aldehydes.

Because insects also express many GPCRs including homologs of human proteins (e.g., serotonin and histamine receptors), it was expected that invertebrate ORs would be readily isolated through homology searches. While this was true for the nematode *Caenorhabditis elegans*, it took until 1999 for the first insect OR to be identified from the vinegar fly (*Drosophila melanogaster*) using unbiased approaches. This is because insect ORs are not GPCRs but an unrelated group of receptor proteins with a similar tertiary structure. Being different to classic GPCRs, the signaling mechanisms have also proven to be different in insects, such as the existence of a highly conserved universal chaperone protein and the activation of both metabotropic and ionotropic signaling cascades (first reported in 2008).

The purpose of this volume is to summarize the latest understanding of molecular mechanisms of olfaction in vertebrates and insects. I have chosen to focus most chapters on insects for several reasons. First, molecular biology of insect olfaction is still an evolving paradigm compared to that of vertebrate olfaction which is relatively well characterized. Second, insects are a megadiverse group that interact with varying levels of specificity, with virtually all other land organisms and therefore as a group have a huge array of ORs that detect countless volatile compounds, many important to humans. This is of great interest in terms of studying general biology but insect ORs also show huge promise in many applications such as pest/disease management and biosensing. Lastly, a lean toward insects gives a point of differentiation with other works on olfaction that have traditionally focused on mammals, of which there are relatively few species.

This first edition of *Molecular Basis of Olfaction* is designed to provide insight into key areas of olfaction research and is intended for use by researchers, teachers, students, molecular biologists, and biologists in general. Leading researchers from China, United States, France, Germany, Sweden, and New Zealand have contributed the chapters presented here, and I take this opportunity to sincerely thank all authors for their effort and expertise.

The chapter "Mammalian Olfactory Receptors: Molecular Mechanisms of Odorant Detection, 3D-Modeling, and Structure–Activity Relationships" by Persuy and coworkers from France summarizes our knowledge of molecular mechanisms of odorant detection in mammals and includes 3D modeling of mammalian ORs, and relationships between receptor structure and activity. In chapter "Olfactory Signaling in Insects," Dieter Wicher (Max Planck Institute for Chemical Ecology) discusses cellular signaling in various types of olfactory neurons in insects. The chapter "Advances in the Identification and Characterization of Olfactory Receptors in Insects" by Montagné *et al.* provides an insight into the latest advances in isolating and characterizing insect ORs, including the use of transcriptomics. The final two chapters focus on specific areas of insect olfaction research of importance to humans. The chapter "Olfactory Disruption: Toward Controlling Important Insect Vectors of Disease" by Sparks *et al.* (U.S. Department of Agriculture) discusses disruption of olfaction in insect vectors of human disease such as mosquitoes and tsetse flies. The last chapter ("Pheromone Reception in Moths: From Molecules to Behaviors" by Zhang and colleagues) summarizes knowledge of one of the great olfactory phenomena in biology, pheromone detection by moths, and the events leading from antennal detection of a pheromone to neural processing and resultant behaviors.

I anticipate that future editions of this volume will update these summaries as well as expanding the focus of the current edition.

Richard Glatz
19 November 2014
Kangaroo Island, Australia

CHAPTER ONE

Mammalian Olfactory Receptors: Molecular Mechanisms of Odorant Detection, 3D-Modeling, and Structure–Activity Relationships

Marie-Annick Persuy*, Guenhaël Sanz*, Anne Tromelin†, Thierry Thomas-Danguin†, Jean-François Gibrat‡, Edith Pajot-Augy*,[1]

*INRA UR 1197 NeuroBiologie de l'Olfaction, Domaine de Vilvert, Jouy-en-Josas, France
†INRA UMR 1129 Flaveur, Vision et Comportement du Consommateur, Dijon, France
‡INRA UR1077 Mathématique Informatique et Génome, Domaine de Vilvert, Jouy-en-Josas, France
[1]Corresponding author: e-mail address: edith.pajot@jouy.inra.fr

Contents

Abstract

This chapter describes the main characteristics of olfactory receptor (OR) genes of vertebrates, including generation of this large multigenic family and pseudogenization. OR genes are compared in relation to evolution and among species. OR gene structure and selection of a given gene for expression in an olfactory sensory neuron (OSN) are tackled. The specificities of OR proteins, their expression, and their function are presented. The expression of OR proteins in locations other than the nasal cavity is regulated by different mechanisms, and ORs display various additional functions.

A conventional olfactory signal transduction cascade is observed in OSNs, but individual ORs can also mediate different signaling pathways, through the involvement of other molecular partners and depending on the odorant ligand encountered. ORs are engaged in constitutive dimers. Ligand binding induces conformational changes in the

Progress in Molecular Biology and Translational Science, Volume 130
ISSN 1877-1173
http://dx.doi.org/10.1016/bs.pmbts.2014.11.001

ORs that regulate their level of activity depending on odorant dose. When present, odorant binding proteins induce an allosteric modulation of OR activity.

Since no 3D structure of an OR has been yet resolved, modeling has to be performed using the closest G-protein-coupled receptor 3D structures available, to facilitate virtual ligand screening using the models. The study of odorant binding modes and affinities may infer best-bet OR ligands, to be subsequently checked experimentally. The relationship between spatial and steric features of odorants and their activity in terms of perceived odor quality are also fields of research that development of computing tools may enhance.

1. MAMMALIAN OLFACTORY RECEPTORS: FROM GENES TO PROTEINS

Olfactory receptors are predominantly expressed in the main olfactory epithelium located in the nasal cavity. They are the gateways, located across the plasma membranes of olfactory sensory neurons (OSN) cilia, through which the message conveyed by the odorant molecules in the ambient air transit, before being transduced into an electrical signal.

1.1. Genes and pseudogenes

In mammals, there exist several hundred (up to several thousand) OR genes accounting for 1–3% of estimated mammalian gene repertoire,[1,2] and representing the largest gene superfamily.

The number of OR genes exceeds 1700 in the rat and is around 860 in humans.[3] This abundance is justified by the number of physiological functions in which olfaction is involved (food intake and preferences, search for prey, predator avoidance, social behaviors, mother–young relationships, spatial orientation, stress, etc.), even though this chemical sense was for a while considered to be a minor sense relative to vision. ORs being GPCRs are characterized by seven-transmembrane helices (TMHs), participating in the transmission of the olfactory message carried by the volatile odorant compounds of the environment.[4–6] Because ORs are involved in the detection of chemical messages from the environment of animals, their genes have undergone selection pressure, inducing the evolution of the olfactory repertoires of the various species. Some OR genes evolved to nonfunctional pseudogenes[7] in varying proportions depending on the species, from $\approx$20% in the mouse and dog[8,9] to $\approx$50–60% in primates and humans[1,3,10] (for review, see Ref. [11]). Indeed, if the number of OR genes differs from species to species (133 ORs in zebrafish to 1300 in pigs,[12] 2129 in cows, 4200 in African elephants[13]) the amount of pseudogenes is also variable. Some primates have less than 400 types of functional ORs (humans and chimpanzees, orangutans, and macaques even less[14,15]) compared to over 1000 for pigs, rodents and dogs,[12,16,17] and 1948 in African elephants.[13] However, the cognitive power of these species, i.e., the ability to process olfactory data, allows them to integrate information from complex olfactory environments, beyond simply the number of functional ORs that can be activated.[18]

Mammalian OR genes are organized in a large number of clusters distributed on many chromosomes e.g., 9 chromosomes for mice,[19] all chromosomes except 20, and Y for humans.[7] Potentially, coding sequences may predominate on some chromosomes (7, 16, and 17 in humans, for instance[7]). OR pseudogenes are interspersed with full-length OR genes. Closely located OR genes within a cluster tend to be closely related evolutionarily, while duplication of whole OR gene clusters appears to be rare.[20] Generation of this large and diverse multigenic family involved in a key biological function may result from successive duplications of large genomic regions during evolution,[11,21] followed by an accumulation of mutations. Moreover, evolutionarily distantly related genes may be found in a given OR gene cluster, and OR genes with a close evolutionary relationship may be located at different clusters or chromosomes,[20] suggesting additional chromosomal rearrangements within OR gene clusters and shuffling of the genes from different clusters.

In different species, a number of OR genes exhibit sequence identities above 90%, for instance in dogs and humans,[22] humans and other primates,[7,14,23–25] rats and mice.[25] Man *et al.*[26] showed that orthologs (coded by genes deriving from the same ancestor by speciation) were more similar than paralogs (coded by genes deriving from the same ancestor gene by duplication) when measuring amino acid similarity, using either the whole coding sequence or the 22 amino acids predicted to be involved in ligand binding. In closely related species, orthologs tend to present similar ligand selectivity but important differences in receptor potency (EC_{50}) to a given ligand. However, while paralogous ORs within the same species respond to a common ligand only 33% of the time, orthologous ORs respond to a common ligand 82% of the time on average (from 93% for human–chimpanzee orthologs to 83% for human–mouse orthologs).[25] Moreover, the genetic variation in the coding region of OR genes may contribute to the variation in odor perception among individuals.

Mammalian OR genes are divided into two classes. Class I was initially ascribed to fish OR genes for which OR proteins mostly bind hydrophilic odorants (amino acids), while Class II was related to mammalian OR genes with OR proteins binding hydrophobic odorants. In fact, recent studies show that Class I ORs can be subdivided into several groups, among which the α group is proposed to encode ORs specific to airborne odorants, while the δ, ε, ζ, and η group genes appear to primarily detect water-soluble odorants. Only the α group of Class I is present in mammals, together with the Class II genes (which consists only of γ group genes).[27] Fishes encode only Class I genes, of groups δ, ε, ζ, and η, and in amphibians OR genes are found from both Classes (Fig. 1). Interestingly, both in the human and mouse genomes, all Class I OR genes (thus of the α group) are encoded in a single genomic cluster, contrary to Class II genes.[11,28] Pseudogenes are present in a lower proportion among human Class I ORs (52%) than Class II ORs (77%),[1] suggesting that "fish" OR genes still have a functional significance.

OR genes exhibit a relatively well-conserved structure including one or several small untranslated exons at their 5′ termini, followed by a large 3–10 kb intron preceding a single coding exon of about 1 kb and a polyadenylation signal.[30] Cloning OR coding sequences from genomic DNA is therefore quite straightforward. The generation of the repertoire of OR genes exhibiting a single coding exon may partly arise from retroposition of OR mRNA in an early evolutionary process.[31] OR gene clusters could have resulted from duplication of these ancestral retrogenes.

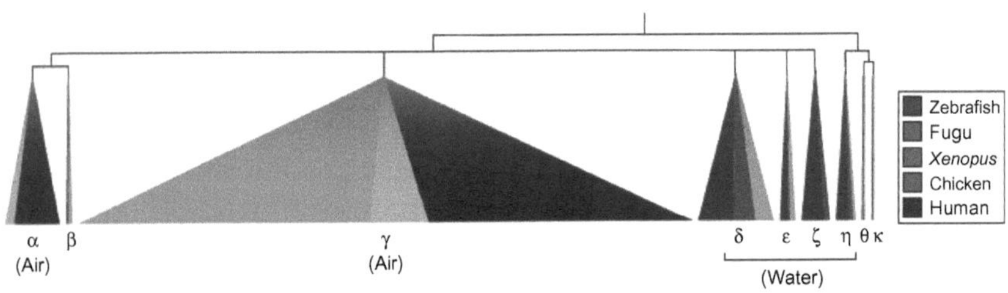

Figure 1 Evolutionary dynamics of OR genes: a phylogenetic tree of OR genes from five vertebrate species. The genes that belong to different groups are represented by different colored triangles. The size of each triangle is approximately proportional to the number of OR genes from each species. The α and γ group genes are proposed to primarily detect airborne odorants because they exist in tetrapods, whereas the δ, ε, ζ, and η group genes that exist in fishes and Xenopus appear to primarily detect water-soluble odorants. The functions of the group β, θ, and κ genes are unclear. *Adapted by permission from Macmillan Publishers Ltd. Nature Reviews Genetics, Ref.* [29] *copyright 2008.* (See the color plate.)

Promoter sequences present a low homology, even for closely related ORs.[2,32,33] An extremely high level of single nucleotide polymorphism is reported in OR promoters, which may be related to personalized odor coding.[34] TATA boxes are found in at least a subset of OR promoters,[35,36] contrary to suggestions from previous studies.[31,32,37] The OR gene transcription efficacy also depends on transcription start sites, which are investigated by large-scale mapping technologies.[38] There does not seem to be a consensus on their location, which still needs to be confirmed. Ongoing studies have shown, on OR gene promoters, an enrichment of binding sites for transcription factors of the O/E family, or for homeodomain factors.[35,39] In addition to a minimal promoter, long-range elements like the so-called core-H (noncoding) region have been shown to regulate expression of all OR genes in the same cluster.[40]

1.2. OR protein expression

OR genes encode integral membrane proteins belonging to the seven-transmembrane domain, GPCR superfamily, participating in the cellular response to environmental chemosensory signals.[4] According to the GRAFS (Glutamate, Rhodopsin, Adhesion, Frizzled/Taste2, Secretin) classification, GPCRs are divided into five families,[41] and all ORs belong to the "rhodopsin-like" receptors or "Class R" family. ORs account for more than half the GPCRs in mammalian species. However, they often exhibit very low sequence identity between each other, except for some characteristic consensus sequences.[6] ORs seem to carry no signal peptide sequence. Their N-terminal end is extracellular and short, while the C-terminal part is intracellular and interacts with the G-proteins.

OR expression was first discovered in the olfactory epithelium by Buck and Axel,[4] who were later awarded the Nobel prize for this. ORs are located at the membrane of the dendrites of OSNs, and each OSN expresses a single allele of a single OR gene. The spatial organization of these genes in the chromatin of a given neuron is likely to be important for both the monoallelic and monogenic character of their expression.[42] The OR choice seems to involve an escape from silencing, in a model in which all OR genes in olfactory neuron progenitors initially reside in inactive heterochromatin, and derepression of a given gene by demethylation of the repressive histone marker H3K9me3, allows its expression.[43–45] The two homologous alleles of a given OR gene are associated with different heterochromatin domains; one with deeply repressed constitutive heterochromatin and thus permanently repressed, the other one with the more plastic facultative heterochromatin, thus available for transcription.[46] A Locus Control Region located upstream of OR genes (the so-called core-H region), to which chromatin-remodeling/transcription-activating factors can bind, physically interacts with one promoter site through random collision, thereby remodeling the chromatin structure, and activating one particular OR gene within the cluster[47,48] (for review, see Ref.[49]). The facultative heterochromatin domains could themselves result from the negative feedback signal elicited by an expressed OR gene to prevent the expression of additional ORs, thereby contributing to the stability of OSN OR gene choice.[50,51] In fact, once an OR gene is activated, its expression may inhibit further activation of other OR genes by downregulating a histone demethylase required for the removal of the repressive histone marker H3K9me3 on OR genes, which would allow their expression.[52] However, the presence of transcripts for two different ORs was reported in a subset of OSNs,[53] possibly resulting in the coexpression of these two ORs.

The expression of most OR genes of Class I appears to be confined to the dorsal region in the mouse olfactory epithelium.[5,54,55] This is in line with the presence of common sequences in their promoters that may restrict their expression to specific regions.[39] As for Class II OR genes, their expression in OSNs is scattered in partially overlapping regions of the epithelium.[55] This suggests that their expression pattern may arise from gene-specific promoters.[39]

The OR role in the olfactory epithelium is to detect and discriminate odorant molecules according to a combinatorial code in which an OR can detect various odorant molecules and an odorant can activate various ORs. Thus, a mixture of odorants activates a specific group of ORs and there may be some overlapping between the groups of ORs stimulated by different odorants.

Besides their well-known role in odorant detection from the air inspired through the nose, ORs appear to exhibit additional functions when expressed in locations other than the ciliae of the OSNs (for review, see Ref.[56–58]). ORs may be locally synthesized in OSN axons emerging from the olfactory epithelium[59] and contribute to axon sorting by favoring and stabilizing fascicles of axons expressing the same OR,[60] in a model where both homo- and heterotypic dynamic axon–axon interactions may mediate adhesion.[61] Pronin *et al.*[62] reported the expression of an OR in arterioles of the eye, suggesting a role in sensing chemicals in its environment. Some ORs are involved in sperm chemotaxis and migration,[63–66] and in cell migration and adhesion in the skeletal muscle.[67] ORs expressed in the kidney may modulate renin secretion and regulate blood pressure,[68,69] and ORs in enterochromaffin cells induce serotonin secretion in the gut.[70,71] Several ORs were also reported in duodenal enterocytes, some of them being upregulated by a high-fat diet in obesity-prone rats. These receptors may thus be involved in the regulation of dietary fat, and in individual susceptibility to obesity.[72] Eleven ORs of Class II were also found in rat placenta.[73] However, most studies only demonstrate OR transcript presence,[58] with no evidence of protein expression.

A specific OR expression was detected in both primary small intestine neuroendocrine carcinoma and metastases, and could thus constitute a potential novel clinical tissue biomarker.[74] Other ORs are also reported to be overexpressed in tumor cells where they constitute tumor markers. This upregulation should be explored more extensively, since ORs could be involved in tumor progression.[75–78] ORs are reported to participate in early cytokinesis by exerting a regulatory role on the actin cystoskeleton, and particularly in cancer cell lines.[79]

The regulation of OR gene expression seems to be different in OSNs compared to other cells. Indeed, it was reported that sperm cells and enterochromaffin cells coexpress various ORs contrary to OSNs.[63,78] Eight OR transcripts were detected in pulmonary macrophages (and OR protein presence was confirmed for one of them), with a potential role in the response to microbial infection which seems to be mediated by bacteria-released odorants promoting macrophage migration and accumulation at the site of infection.[80]

1.3. Olfactory signal transduction

Events resulting from odorant binding on ORs and subsequent triggering of the olfactory signal remain poorly known. Indeed *in vitro* expression of functional ORs at a significant level is still a challenge for investigating the mechanisms involved. This results from a poor trafficking of the receptors to the plasma membrane in heterologous systems, although expression was performed in various systems, including bacteria, yeasts, insect cells, *Xenopus* oocytes, and mammalian cells, some possibly derived from olfactory epithelium. This also partly explains the still high percentage of orphan ORs: only 8% and 10% of the mouse and human ORs repertoires, respectively, have been deorphaned, i.e., at least some of their ligands have been identified, as of the beginning of 2014[81,82] (for review, see Ref.[83]). It has been shown that some GPCRs require dimerization or association to chaperone proteins for adequate folding and membrane targeting. Similarly, a number of studies have shown that ORs exist as dimers with other GPCRs (adrenergic, purinergic, or adenosine receptors),[84,85] or are associated with other membrane proteins (receptor expression enhancing protein and receptor transporting protein).[86] However, this cannot yet be extended to all ORs.

Although ORs mediate various functions depending on their expression site, the signal transduction cascade is mainly described in the OSNs of the olfactory epithelium. ORs are expressed at the surface of the ciliae that emerge from the dendritic knob of the OSNs into the nasal cavity and are bathed by the olfactory mucus. Olfactory transduction covers all the biochemical (production of second messengers) and electrical (opening of ionic channels) steps from odorant-ligand binding on the OR until the emission of action potentials by the OSN. In mammals, the majority of OSNs of the olfactory epithelium share the same signaling pathway in the olfactory ciliae, where all proteic actors are present.[87–91] Binding of a ligand to an OR activates a heterotrimeric G-protein composed of a GTP-binding $G\alpha_{olf}$ protein subunit and of a $\beta\gamma$ dimeric complex.[92] $G\alpha_{olf}$ dissociates from the $\beta\gamma$ complex upon GTP binding, and selectively stimulates the adenylate cyclase III enzyme, responsible for cAMP (cyclic adenosine 3′,5′-monophosphate) synthesis. In mouse OSNs, $G\beta_1$ and $G\gamma_{13}$ seem to be the exclusive $\beta\gamma$ partners of $G\alpha_{olf.}$[93] cAMP acts as second messenger, by activating the opening of cyclic nucleotide-gated channels, which results in the inward flow of the mainly extracellular Na^+ and Ca^{2+} cations.[94] In turn, the increase of Ca^{2+} concentration in the olfactory ciliae opens the Cl^- channels, inducing an outward flow of Cl^-, which further depolarizes the neuron locally and transiently, resulting in the generation of a receptor potential. The amplitude of this depolarization depends on the nature and amount of the odorant molecules detected by the ORs. The receptor potential triggers an action potential, which is emitted with a frequency depending upon the intensity and duration of the olfactory message. Trains of action potentials (spike trains) are transmitted along the axons of the OSNs toward the olfactory bulb, which is the first integration relay of the olfactory message.

However, other studies reveal that in some OSNs, ORs can mediate different signaling pathways, even when activated by structurally similar ligands.[95] This might be due to the different conformations of the intracellular regions of ORs induced by the binding of different odorants, which have an impact on the selectivity of coupling to the $G\alpha$ proteins.

Depending on the combinations of cellular partners present and on the odorant considered, the stimulation of an OR may orient the response of the OSN toward different signaling pathways, due to the type of Gα protein that is coupled, the effector involved, and the second messengers. Individual ORs can use pathways other than cAMP production to increase intracellular calcium concentration, providing another mode for odorant signaling in the olfactory system.[96] Indeed, the phospholipase C-β2 (PLC-β2) pathway may be activated instead of the adenylate cyclase pathway.[53,97–99] Some studies provide evidence that these pathways do not work independently in rat olfactory neurons, but rather show a functional antagonism.[100] Although the PLC-β2 pathway and its second messenger product IP3 were implicated in odor transduction in fish,[101,102] amphibians,[103] and lobster,[104] the activation of PLC in response to odors may be indirect and constitute a modulation of the odor transduction.[105,106]

Some of the other cell types expressing ORs also express part of the canonical signaling pathway ($G\alpha_{olf}$, possibly adenylate cyclase III).[62,68,70,73] This suggests that the olfactory machinery may be involved in additional functions in other tissues. Moreover, odorant mixtures can induce unpredictable responses, due to possible competitive or additive effects between odorants or signaling pathways.[107–109]

2. OLFACTORY RECEPTOR ACTIVITY REGULATION: HOMODIMERIZATION, BINDING COOPERATIVITY, AND ALLOSTERY

The functional response of some ORs expressed in heterologous systems, such as mammalian cells (e.g., HEK293) or yeasts (e.g., *Saccharomyces cerevisiae*), displays a bell-shaped dose–response curve with increasing odorant doses.[110,111] This appears in apparent contradiction with the sigmoid curves observed by stimulating ORs in natural tissues.[112,113] Yet, a decreased response of ORs at high odorant doses can be explained by a model involving allosteric modulation of OR activity by OBPs[114] (Fig. 2) and ligand binding cooperativity within an OR homodimer.[115] On the one hand, it was described that OBPs can bind ORs[116] and restore OR activity at high odorant doses.[114] OBP modified the functional OR-1740 dose–response to helional, from a bell-shaped to a saturation curve, thus preserving OR activity at high ligand concentration. This unravels an active role for OBPs in olfaction, in addition to a passive transport or scavenger role. It is also consistent with a physiological effect, in which olfactive sensing is kept upon

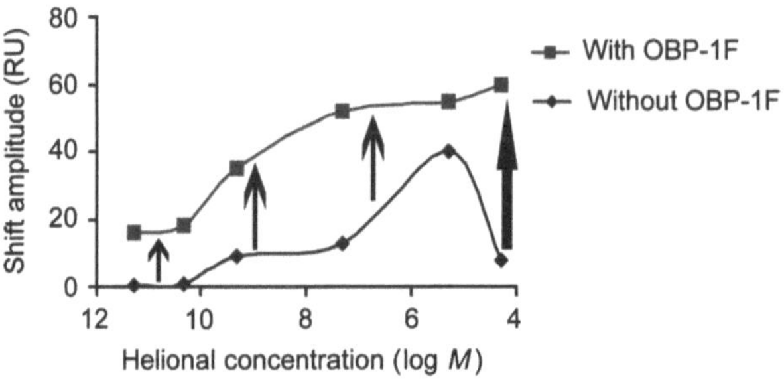

Figure 2 Effect of OBP-1F on helional detection by OR17-40 assayed by surface plasmon resonance (SPR). Each curve is plotted as the difference in response to helional relative to controls obtained by replacing the odorant with water. The SPR shift amplitude is shown as a function of the helional concentration, without or with OBP-1F. The OBP restores OR activity at high odorant doses, changing the response curve from bell-shaped to sigmoidal. *Adapted from Ref.* [114] *with permission from the Royal Society of Chemistry.*

approaching the source of an odorant plume, to maintain an animal's behavioral response toward food or predators for instance. On the other hand, ORs were shown to exist as constitutive homodimers using bioluminescence resonance energy transfer (BRET).[115] Thus, it was assumed that OBPs could regulate OR activity by exerting an allosteric control within OR dimers.

Furthermore, OR dimers were demonstrated to display different conformational changes upon stimulation with various odorant doses, corresponding to different levels of activity[115] (Fig. 3). At low doses, odorants induce a first conformational change in the OR dimers (shown as an increase of the initial BRET level, that is due to the presence of constitutive OR dimers) and are able to activate ORs, whereas at higher doses, odorants induce another conformation of the OR dimers (shown by a smaller increase of the initial BRET level) and are less efficient in activating the receptors. It was thus proposed that at low odorant doses, only one odorant molecule could bind to the OR dimer on one protomer, this binding inducing a conformational change of the second protomer that reduces its affinity for the odorant. OR dimers binding only one odorant molecule would be in an active form. On the contrary, at high ligand doses, the free and low affinity protomer of the OR dimer could bind a second odorant molecule, leading to an inactive conformation of the receptors. Yet, in the presence of OBPs and at high odorant doses, OBPs binding to the OR dimer at an allosteric site would prevent the binding of a second odorant molecule and would thus preserve OR activity. Such a "multistate" model in which the receptor

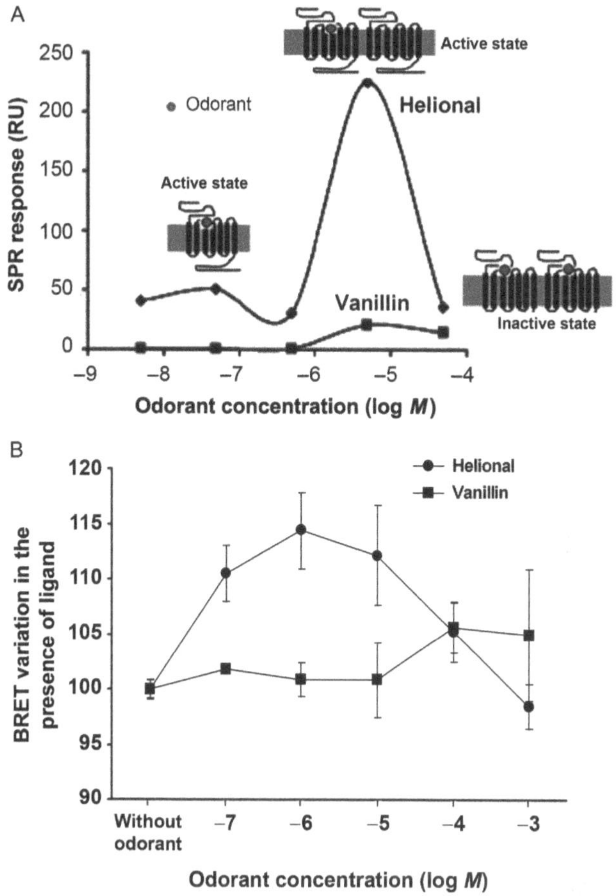

Figure 3 (A) Surface plasmon resonance (SPR) response (RU: relative units) obtained from the stimulation of the OR17-40 receptor with helional (agonist) or vanillin (negative control odorant) at different concentrations. A schematic representation of the proposed molecular mechanism for odorant interaction with the OR is shown. At low and moderate odorant doses, the receptor dimer binds only one odorant molecule and is active, while at high odorant doses it binds two odorant molecules and is in an inactive state. (B) Bioluminescence resonance energy transfer (BRET) level variation upon OR17-40 stimulation with various helional or vanillin concentrations. BRET levels are expressed relative to that measured in the absence of odorant. OR17-40 receptor dimer conformational changes, induced upon stimulation with various odorant doses, elicit an evolution of the BRET level that correlates with the different levels of activity shown in (A). *Panel (A) This figure was originally published in Ref.* [115] *© the American Society for Biochemistry and Molecular Biology.*

activity depends on the occupation rate of the various sites on the dimers has already been reported for other GPCRs.[117]

Since there is increasing evidence that ORs can display pathophysiological functions outside the olfactory epithelium[58,67,69,71,77,79,118] and in particular they can be tumor markers and involved in tumor cell invasion and metastasis emergence;[76,78,119] this negative modulation of OR activity by odorants themselves must be taken into account when aiming to control OR activity in a therapeutic context.

3. OLFACTORY RECEPTOR 3D MODELING AND USE FOR VIRTUAL SCREENING

With the advent of powerful high-throughput sequencing technologies, the so-called next-generation sequencing technologies, the genomes of many organisms have been sequenced and analyzed. Using *in silico* homology search techniques, these analyses have revealed the existence of many OR genes and pseudogenes.[120] Contrasting with this wealth of data available *in silico*, very few OR proteins have been studied experimentally. In particular, the ligands of most ORs are unknown (they are termed orphan ORs). As mentioned previously, ORs can be activated by several ligands (odorants are usually low molecular weight, airborne molecules) and a ligand can activate several ORs. This leads to a combinatorial mechanism that endows organisms with the capability of potentially recognizing ten of thousands of odorants. To help explore the extremely wide range of potential odorant ligands, researchers can rely on computer-aided molecular design techniques that have proven useful in drug design.[121] Two approaches are available: ligand-based techniques, which will be described in the next section, and structure-based techniques. The former requires knowledge of a validated set of ligands with known properties (e.g., agonists, antagonists, inverse agonists). Unfortunately, they are ill adapted to orphan receptors for which, by definition, this information is missing. However, such a ligand-based approach was successfully applied for a human OR for which agonists and antagonists were known.[122] The latter is based on the knowledge of the receptor three-dimensional (3D) structure. This 3D structure is used to perform virtual screening (VS) in which large libraries of chemical compounds are computationally docked to the 3D structure to predict their binding modes and affinities.[123] The 3D structure of receptors can be obtained by biophysical methods (X-ray crystallography or NMR spectroscopy) or by molecular modeling techniques if the 3D structure of a sufficiently close homolog is known.

Until now, no OR 3D structure has been experimentally determined. However, as mentioned above, ORs belong to the large GPCR superfamily. According to the GRAFS classification,[41] ORs belong to the δ-subclass of the R (rhodopsin-like) class, with which they form a monophyletic cluster in phylogenetic analyses. During the last 4 years, an increasing number of 3D structures of the R class have been solved. Table 1 displays the 21 different R class receptors for which 3D structures have been solved so far (and the corresponding literature), including two receptors of the δ-subclass that should be the closest relatives of the ORs (the P2Y purinoreceptor 12 and the human protease activated receptor). As shown in Table 1, GPCRs have been crystallized when bound to different ligand types (inverse agonists, antagonists, partial agonists, agonists—including some endogenous ones, biased agonists) resulting in the resolution of different conformational states for the receptors. Receptors crystallized with inverse agonists, antagonists, or partial agonists are in an inactive conformational state, and those crystallized with agonists are in a partially active conformation. Only ternary complexes composed of the receptors, agonists, and the whole G-protein heterotrimer[134] (or a camelid nanobody, which mimics the behavior of the α-subunit of the G-protein, in other structures) have been successfully utilised to produce activated conformations for structural resolution. Indeed, experimental evidence shows that G-proteins are necessary to fully stabilize the GPCR activated conformation,[164] with the exception of rhodopsin whose covalently bound ligand (the retinal that switches from a *cis* to a *trans* conformation upon being hit by a photon) appears sufficient to stabilize the activated conformation.[165] The knowledge of these structures exhibiting activated receptor conformation combined with biophysical techniques such as NMR spectroscopy,[166] helped researchers to gain more insight into the molecular basis of the signal transduction mechanism.[167,168]

This wealth of experimental data has allowed a better sampling of the GPCR families, subfamilies and subtypes; besides R class receptors, structures of GPCRs from the S class (Secretin-like),[169] F class (Frizzled-like),[170] and G class (Glutamate-like)[171] have also been solved recently. It has also provided a structural framework to understand GPCR activation: large-scale rearrangement of TMHs, and identification of residues acting as ligand-dependent "triggers" and conserved microswitches in these helices.[172]

All this information can be advantageously harnessed to discover new OR ligands using VS techniques.[173–176] Since no 3D structure of ORs

Table 1 Experimental structures of complexes of GPCRs with different types of ligand

Receptor	Class/subclass[a]	Year	PDB code	Resolution (Å)	Ligand	Type	Conformational state	References
Bovine rhodopsin	R/α	2000	1F88	2.8	11-*cis* Retinal	Inverse agonist	Inactive	Palczewski *et al.*[124]
		2004	1U19	2.2	11-*cis* Retinal	Inverse agonist	Inactive	Okada *et al.*[125]
		2011	3PQR	2.8	All-*trans* retinal	Agonist	*Activated*[1]	Choe *et al.*[126]
		2011	3OXA	2.6	Beta-ionone	*Allosteric site*[b]	Inactive	Makino *et al.*[127]
Squid rhodopsin	R/α	2008	2Z73	2.5	11-*cis* Retinal	Inverse agonist	Inactive	Murakami and Kouyama[128]
Human β2 adrenergic	R/α	2007	2RH1	2.4	Carazolol	Inverse agonist	Inactive	Rasmussen *et al.*[129]
		2008	3D4S	2.8	Timolol	Inverse agonist	Inactive	Hanson *et al.*[130]
		2010	3NY8	2.8	ICl118551	Inverse agonist	Inactive	Wacker *et al.*[131]
		2010	3NY	3.2	Alprenolol	Antagonist	Inactive	Wacker *et al.*[131]
		2011	3P0G	3.5	BI-167107	Agonist	*Activated*[2]	Rasmussen *et al.*[132]
		2011	3PDS	3.5	FAUC50	Irreversible agonist	Intermediate	Rosenbaum *et al.*[133]
		2011	3SN6	3.2	BI-167107	Agonist	*Activated*[3]	Rasmussen *et al.*[134]
		2013	4LDL	3.1	Hydroxybenzylisoproterenol	Agonist (high affinity)	*Activated*[2]	Ring *et al.*[135]

		2013	4LDO	3.2	Adrenaline	Agonist (endogenous)	*Activated*[2]	Ring *et al.*[135]
		2013	4LDE	2.8	BI-167107	Agonist (ultrahigh affinity)	*Activated*[2]	Ring *et al.*[135]
Turkey β1 adrenergic	R/α	2008	2VT4	2.7	Cyanopindolol	Antagonist	Inactive	Warne *et al.*[136]
		2011	2Y00	2.5	Dobutamine	Partial agonist	Inactive	Warne *et al.*[137]
		2011	2Y04	3.0	Salbutamol	Partial agonist	Inactive	Warne *et al.*[137]
		2011	2Y02	2.6	Carmoterol	Agonist	Inactive	Warne *et al.*[137]
		2011	2Y03	2.8	Isoprenaline	Agonist	Inactive	Warne *et al.*[137]
		2011	2YCW	3.0	Carazolol	Inverse agonist	Inactive	Moukhametzianov *et al.*[138]
		2012	4AMI	3.2	Bucindolol	Biased agonist	Inactive	Warne *et al.*[139]
		2012	4AMJ	2.3	Carvedilol	Biased agonist	Inactive	Warne *et al.*[139]
		2013	3ZPR	2.7	Quinoline	*Possibly an antagonist*	Inactive	Christopher *et al.*[140]
Human A2A adenosine	R/α	2008	3EML	2.6	ZM241385	Inverse agonist	Inactive	Jaakola *et al.*[141]
		2011	3QAK	2.7	UK-432097	Agonist	Intermediate	Xu *et al.*[142]

Continued

Table 1 Experimental structures of complexes of GPCRs with different types of ligand—cont'd

Receptor	Class/ subclass	Year	PDB code	Resolution (Å)	Ligand	Type	Conformational state	References
		2011	2YDO	3.0	Adenosine	Agonist	Intermediate	Lebon *et al.*[143]
		2011	2YDV	2.6	NECA	Agonist	Intermediate	Lebon *et al.*[143]
		2011	3REY	3.3	XAC	Antagonist	Inactive	Dore *et al.*[144]
		2011	3RFM	3.6	Caffeine	Antagonist	Inactive	Dore *et al.*[144]
Human chemokine CXCR4	R/γ	2010	3ODU	2.5	IT1t	Antagonist	Inactive	Wu *et al.*[145]
		2010	2OEO	2.9	CVX15	Antagonist	Inactive	Wu *et al.*[145]
Human dopamine D3	R/α	2010	3PBL	2.9	Eticlopride	Antagonist	Inactive	Chien *et al.*[146]
Human histamine H1	R/α	2011	3RZE	3.1	Doxepin	Inverse agonist	Inactive	Shimamura *et al.*[147]
Human sphingosine 1-phosphate	R/α	2012	3V2Y	2.8	ML056	Antagonist	Inactive	Hanson *et al.*[148]
Human M2 muscarinic acetylcholine	R/α	2012	3UON	3.0	QNB	Antagonist	Inactive	Haga *et al.*[149]
		2013	4MQS	3.5	Iperoxo	Agonist	*Activated*[2]	Kruse *et al.*[150]
Rat M3 muscarinic acetylcholine	R/α	2012	4DAJ	3.4	Tiotropium	Inverse agonist	Inactive	Kruse *et al.*[151]
Mouse μ-opioid	R/γ	2012	4DKL	2.8	β-FNA	Irreversible antagonist	Inactive	Manglik *et al.*[152]
Human κ-opioid	R/γ	2012	4DJH	2.9	JDTic	Selective antagonist	Inactive	Wu *et al.*[153]

Mouse δ-opioid	R/γ	2012	4EJ4	3.4	Naltrindole	Selective antagonist	Inactive	Granier *et al.*[154]
Human nociceptin/ orphanin FQ	R/γ	2012	4EA3	3.0	C-24	Antagonist	Inactive	Thompson *et al.*[155]
Rat neurotensin	R/β	2012	4GRV	2.8	Neurotensin	Agonist	Intermediate	White *et al.*[156]
Human protease activated receptor 1	R/δ	2012	3VW7	2.2	Vorapaxar	Antagonist	Inactive	Zhang *et al.*[157]
Human chemokine CXCR1	R/γ	2012	2LNL	NMR	–	–	Inactive	Park *et al.*[158]
5-Hydroxytryptamine (5-HT) receptor	R/α	2013	4IBA	2.7	Ergotamine	Biased agonist	Intermediate	Wacker *et al.*[159]
CCR5 chemokine receptor	R/γ	2013	4MBS	2.7	Maraviroc	Inverse agonist	Inactive	Tan *et al.*[160]
P2Y purinoceptor 12	R/δ	2014	4PXZ	2.5	2MeSADP	Agonist	Inactive	Zhang *et al.*[161]
		2014	4NTJ	2.6	AZD1283	Reversible antagonist	Inactive	Zhang *et al.*[162]
Free fatty acid receptor 1	R/?	2014	4PHU	2.3	TAK-875	Partial agonist	Inactive	Srivastava *s.*[163]

[a]GRAFS classification.[41]
[b]Beta-ionone binds to an allosteric site; ternary complex with: [1]11 amino-acid C-terminal fragment of Gt α-subunit, [2]camelid nanobody, [3]Gs protein.

has been elucidated yet, the first step of an alternative approach is to build a model of the 3D structure of the ORs of interest using molecular homology modeling techniques.[177] The second step consists of docking, *in silico*, a library of small molecules with these 3D structures to find those that exhibit the best predicted affinity with the corresponding receptors.[123]

3.1. Model building

Before the first high-resolution GPCR structure was available, that of rhodopsin in 2000,[124] GPCR models, including some early OR models, were based on *de novo* modeling of the GPCR characteristic 7-TMHs.[178] With the availability of an increasing number of new, high-resolution GPCR 3D structures, it is becoming increasingly beneficial to use homology modeling techniques based on these structures, rather than *de novo* modeling. Homology modeling is based on the notion of homology, which is a central concept in biology. Two genes are homologous if they descend from a common ancestor. The product of this ancestor (the ancestor protein) had a particular sequence, 3D structure, and function. Its modern descendants *may* have retained similar sequences, have kept the same global 3D structure, and often exhibit closely related functions. The point that concerns us here is the fact that homologous proteins have kept very similar global 3D structures. Therefore, it is possible to build a model of the structure of a protein from the knowledge of the 3D structure of one of its homologs. There is a correlation between the sequence similarity of the two homologous proteins and the resemblance of their 3D structures. As a rule of thumb, when the sequence identity (computed after having aligned the two sequences) is above 50%, the two corresponding structures are very similar (the difference is within experimental errors). Below this value, even though the two proteins retain the same fold (global 3D structure), details of the 3D structure start to differ increasingly (for instance the secondary structure elements move a few Angstroms relative to each other). The smaller the sequence identity, the larger the 3D structure differences. The largest structural differences are observed in the loop regions that are usually less conserved.

Homology modeling techniques consist of four steps: (i) search for a template (the protein to be modeled is called the *query* and the homologous protein whose 3D structure is known is called the *template*), (ii) alignment of the query and template sequences, (iii) construction of the query 3D model *based* on this alignment, and (iv) validation of the query model. ORs belong to the GPCR superfamily, therefore, step (i) is a formality. Step (ii) is absolutely crucial. If the alignment is faulty, the resulting model will be irremediably erroneous. For instance, for GPCRs, a shift of two residues in the TMHs of the model will cause residues that ought to be in the lumen of the binding pocket to face the membrane on the opposite face of the helix with, quite obviously, disastrous consequences in the subsequent docking stage. The alignment of ORs with the 21 GPCRs whose structure has been solved is a tricky point since the sequence identity between OR and other GPCRs is often below 20%. In many GPCR sequence alignments, this difficulty is mitigated by the good conservation, within the TMHs, of a number of sequence motifs or residues (for instance, the ones that are at the basis of the Ballesteros–Weinstein numbering scheme[179]). These motifs and residues help anchoring the alignments of the TMHs. Unfortunately, Ballesteros–Weinstein N50 residues of TMH5 and TMH6 are not conserved in most OR sequences. N50 is the most conserved residue in each TMH as observed in GPCR multiple sequence alignments. However, notice that Ballesteros and Weinstein did not include ORs in their alignments.[179] Likewise, the CWxP microswitch motif of TMH6 is not well conserved. Therefore, accurately aligning TMH6 can be challenging. Regarding step (iii), the most structurally conserved region in GPCRs is the 7-TMH domain (7TMD). Comparisons of crystal structures listed in Table 1 show that 7TMDs are sufficiently similar in known GPCRs to form a good basis for building their counterparts in query proteins. It is much more challenging to accurately model the three extracellular and three intracellular loops. Loops are often the most variable regions in proteins and, indeed, this is what is observed when analyzing the known GPCR structures. One must thus resort to *de novo* or knowledge-based loop modeling.[180] Even with these techniques, accurately modeling loops (i.e., with a root-mean-square deviation of the Cα atoms less than 2 Å) longer than 12 residues still remains a difficult task.[181] The second extracellular loop (ECL2) has been shown to be important for ligand binding in some of the GPCRs listed in Table 1. Its length can be up to 30 residues in many GPCRs (this is the case for ORs), and it is thus difficult to model precisely. Possibly, the resulting model can be optimized using different techniques described in Ref.[182] Finally, the fourth and last step is model validation, whereby both theoretical evaluations (employing tools for estimating the correctness of crystallographic structures) and available experimental validations (known mutations, cysteine accessibility, structure–activity relationship information) can be used. To give maximum confidence in the resulting models, it is essential to incorporate all the experimental pieces of information at hand about the receptor of interest, into the model.[174] However, the most convincing model validation involves first the prediction of a ligand with a good affinity for the modeled receptor (see Section 3.2) and the identification of the residues that are associated with the ligand binding, *in silico*. Then, one must experimentally (i) verify the ability of this ligand to bind to the receptor or to activate or inactivate it, through functional assays utilizing calcium imaging, electrophysiology, or surface plasmon resonance,[183] and (ii) check that mutating the residues predicted to be important for the ligand binding indeed affects the receptor functional activity.

3.2. Ligand virtual screening

Ligand virtual screening consists of identifying *in silico* from a large library of chemical compounds those that exhibit a good affinity for the receptor. This requires the ability to correctly dock the chemical compounds to the receptor, i.e., to accurately determinate their binding mode, and to faithfully estimate their binding affinity. Physically, ligand affinities depend on the corresponding binding free energies. Accurately computing free energy is a complex endeavor, often resulting in data with rather large errors (up to 1–2 kcal). However, going from a compound having an activity at μM concentration to one having an activity at nM concentration (i.e., three orders of magnitude less) only involves a mere 5 kcal increase in their binding free energies.[184] In addition, methods for accurately computing free energies are computer intensive and cannot easily be applied to the millions of compounds found in the largest libraries. Therefore, most *in silico* docking methods are based on a trade-off between accuracy and speed.

Docking programs consist of two components: (i) a method to explore the conformational space and enumerate binding modes and (ii) a score function to evaluate and rank the proposed binding modes.

Several types of conformational sampling methods have been proposed. In some methods, single or multi-conformer compound libraries are rigidly docked to the receptor, often using shape complementarity. These methods are fast but are not very accurate as the ligand conformations are not sufficiently sampled. Other methods use incremental construction in which the ligand is built up in the binding site from fragments in a stepwise fashion, considering both preferred conformations and ligand flexibility to connect the fragments. However, most methods rely on a stochastic exploration of the conformational space using techniques based on Monte Carlo or genetic algorithms.

The score function provides a more or less accurate proxy for the binding free energy. Docking methods employ three types of score functions: empirical, knowledge-based and force field-based. Empirical score functions are comprised of a number of weighted terms, each describing a particular ligand–protein molecular interaction (e.g., hydrogen bonds, hydrophobic effects, electrostatic effects, etc.), whose weight factors are estimated by regression to fit experimentally determined protein–ligand affinities. Knowledge-based score functions are estimated from a statistical analysis of the observed distances between relevant ligand and protein sites (e.g., hydrogen bond donor and acceptor sites, positively and negatively charged sites, etc.) in known complexes using the relation:

$$\Delta G_X = -\log P_X,$$

where ΔG_X is the free energy contribution of the X type of interaction between two sites and P_X is the probability of X for a particular distance between these two sites.

Force field-based (or physics-based) score functions were originally developed for computing thermodynamic and kinetic properties of small molecules and macromolecules. They incorporate many physical interactions such as: van der Waals, electrostatic, hydrogen bond, and solvation. Physics-based score functions are the most accurate but also the most demanding in terms of computing resources. Moreover, they are less forgiving of small inaccuracies existing in the receptor 3D structure than the two former score functions. Therefore, one must allow the receptor conformation to be flexible in the docking simulations.

Several reviews have been devoted to the detailed evaluation of 3D model building and ligand virtual screening results, some addressing general protein targets[123] and some focusing on GPCRs,[185–187] that the interested reader might benefit from referring to. With regard to OR virtual screening, a list of homology modeling and odorant docking computer simulations that have been carried out since 1994 to decipher the yet largely unexplored odor repertoire of these receptors, is provided in Ref. [178]

3.3. GPCR inverse agonist, antagonist, and agonist ligands

A distinction must be made between the affinity which, as described above, measures the strength of the receptor–ligand interaction and the efficacy that corresponds to the relative ability of a drug–receptor complex to produce a maximum functional response. This efficacy is measured with respect to the endogenous ligand functional response (ELFR) and ranges from full inverse agonist (-100% ELFR) to full agonist (100% ELFR) including partial inverse agonist (-100% < ELFR < 0%), silent antagonist (0% ELFR), and partial agonist (0% < ELFR < 100%). The ligand potency that is often measured by the EC_{50} depends both on the affinity and the efficacy.

Table 1 displays a list of ligands with different efficacies. Notice that biased agonists refer to ligands that do not trigger the canonical downstream signal pathway through binding to a G-protein, but activate other pathways, for instance by binding to β-arrestin or directly through receptor-associated kinases.[176] When performing ligand virtual screening, one is often interested in inhibiting or activating the receptor, i.e., in designing ligands that are either antagonists/inverse agonists or agonists. It is also interesting to conceive biased agonists. Obviously, the affinity does not provide any information about the efficacy. For instance, it is likely that antagonists have a better affinity for the receptor than agonists since they successfully compete with the latter, but their efficacy in stimulating the receptor-mediated signaling is nil.

To design ligands with specific efficacy, it is thus necessary to analyze thoroughly the docking modes and identify which positions in the receptor 3D structures are interacting with these ligands. A chemogenomic analysis of GPCRs[188] and studies of the available 3D structures of the complexes containing different types of ligand have identified a number of positions in the 7TMD and ECL2 that are potentially important to modulate the ligand–receptor efficacy.[159,164,165,167,168,176,189] Often, they correspond to well-conserved sequence motifs such as microswitches and "triggers," as mentioned above. A systematic mapping of these positions and of how they interact with different ligands might help defining whether these ligands are agonists, antagonists, or inverse agonists. In addition to the analysis of X-ray data that provides static picture, biophysical, and computational techniques that allow researchers to study the dynamic behavior of the receptor might help in unraveling this complex issue.[166,190] For one of the ORs studied, based on the analysis of the docking conformations, and supported by receptor mutagenesis and functional assays in a heterologous expression system, authors suggested that antagonists tend to dock in the upper part of the binding pocket whereas agonists dock in the narrow lower part.[191] Of course, further studies are needed to validate or reject this hypothesis for other ORs.

Certainly, with the current wealth of X-ray structures that are increasingly available, investigations of OR properties will develop steadily. To help with this exciting enterprise, an automatic pipeline was developed that allows users to perform OR homology modeling and ligand docking (http://genome.jouy.inra.fr/GPCRautomodel).

4. ODORANT LIGANDS STRUCTURE–ACTIVITY RELATIONSHIPS

Even though the relationship between molecular structure and biological activity has been the focus of researches for decades (see Ref. [192] and cited references), the link between the molecular structure of an odorant and its perceived odor quality was pointed out by Linus Pauling only in the middle of the twentieth century.[193] Nevertheless, the notion of structure–odor relationships was first applied to odorant molecules by Amoore,[194,195] who established a list of primary odors as the results of studies of specific anosmia for different odorants' chemical structures. The development of computational tools further led to the emergence of quantitative structure–activity relationships (QSAR),[196] which attempts to correlate an experimental response (e.g., biological activity or a physicochemical property) with some molecular properties. Following this approach, an experimental response (e.g., odor quality) can be expressed as a function of molecular properties (or molecular descriptors). Several studies have sought to apply the fundamental principle of QSAR to link the chemical structure of odorants to their odor.[197,198] However, establishing reliable structure–odor relationships for the olfactory space as a whole, still remains a desired goal,[199] not yet reached.[200]

Regardless, the identification of a link between the structures of a set of molecules and odor characteristics has proven to be relevant in several approaches applied to air quality, or food and beverage aroma perception,[201–206] which led to the development of QSAR regression tools to obtain predictive models.[203,207–212] Two main odor characteristics have been considered: odor threshold (or intensity) and odor quality (i.e., description). Some strategies have been developed using classification approaches with large sets of odorants,[213–219] while others were more precisely applied to specific molecular structures,[220–222] odor types,[208,223,224] or even hedonic dimensions.[225] The QSAR approach has been largely used to identify the link between odor and molecular structure with the objective of proposing new molecules sharing a peculiar odor.[226,227] Thus, molecular descriptors were used to derive QSAR equations on sets of molecules selected on the basis of odor quality amongst which were defined fruity notes,[208,228] camphor,[208,229] minty,[230] musk notes,[220,221,231] sandalwood,[232–239] and ambergris odors.[233,237,238,240,241]

Many models succeeded in establishing structure–odor relationships for the most rigid molecules such as a-campholenic derivatives (sandalwood odor), trans-decaline (ambergris odor), tetralin, and indans (musk odor). However, the common difficulty for these classical 2D-QSAR experiments was finding a strong relationship for less rigid molecules as for instance, aliphatic esters responsible for fruity notes. Interestingly, this difficulty emphasizes the significance of spatial and steric features. For instance, specific spatial organizations have been reported for sandalwood,[242] bell pepper,[243] and musk[231] odors. Hence, pharmacophore approaches (see Ref. [244] and cited references) combined with the increasing knowledge on ORs constitutes a very promising way to identify the main odorant features responsible for OR activation and perceived odor quality.[245–248]

The pharmacophore approach has contributed to evidence that odorants could be agonists or antagonists of a given OR. This dual agonist–antagonist behavior is likely to be very important in the coding of natural odors that relies on the processing of complex mixtures of odorants. For instance, 3D-quantitative structure–activity relationship (3D-QSAR) was applied to build a double-alignment model explaining *in vitro* experimental activities of a large set of ligands of the human OR1G1 receptor. This approach also successfully predicted new agonists and antagonists for this OR.[122] Moreover, such an approach, associated to the perceived odor quality of these ligands, demonstrated that OR1G1 recognizes a group of odorants that share both 3D structural and perceptual qualities. These results have led to the hypothesis that OR1G1 contributes to the coding of waxy, fatty, and rose odors in humans. These quantitative 3D models remain however very sparse, and few have been reported to date. A previous 3D-QSAR study has been performed that considers an odor as an activity and uses human olfactory detection threshold values as quantitative activity values.[249] The authors of this study used two training sets: the first one was built with nine pyrazines sharing a green odor; the second was built with 10 sweet compounds having various structures (without a common ring structure, unlike the pyrazine derivatives) plus two compounds without sweet odor. Good correlations between steric and electrostatic features of odorants and human olfactory detection threshold values were found for each of the two sets. The authors have highlighted the important role of the size and shape of odorants, assuming that it is related to the direct interactions with ORs, but more complex interactions are probably involved.

The development of computing tools raises the hope for real progress in the knowledge of the chemical space of odorants by chemoinformatics and chemogenomics approaches.[250–253]

REFERENCES

1. Glusman G, Yanai I, Rubin I, Lancet D. The complete human olfactory subgenome. *Genome Res.* 2001;11:685–902.
2. Reed R. After the holy grail: establishing a molecular basis for mammalian olfaction. *Cell.* 2004;116:329–336.
3. Gilad Y, Man O, Glusman G. A comparison of the human and chimpanzee olfactory receptor gene repertoires. *Genome Res.* 2005;15(2):224–230.
4. Buck L, Axel R. A novel multigene family may encode odorant receptors: a molecular basis for odor recognition. *Cell.* 1991;65:175–187.
5. Malnic B, Hirono J, Sato T, Buck L. Combinatorial receptor codes for odors. *Cell.* 1999;96(5):713–723.
6. Gloriam D, Fredriksson R, Schiöth H. The G protein-coupled receptor subset of the rat genome. *BMC Genomics.* 2007;8:338.
7. Rouquier S, Taviaux S, Trask B, et al. Distribution of olfactory receptor genes in the human genome. *Nat Genet.* 1998;18:243–250.
8. Young JM, Trask BJ. The sense of smell: genomics of vertebrate odorant receptors. *Hum Mol Genet.* 2002;11(10):1153–1160.
9. Quignon P, Kirkness E, Cadieu E, et al. Comparison of the canine and human olfactory receptor gene repertoires. *Genome Biol.* 2003;4:R80.81–R80.89.
10. Zozulya S, Echeverri F, Nguyen T. The human olfactory receptor repertoire. *Genome Biol.* 2001;2(6):0018.0011–0018.0012.
11. Niimura Y. Olfactory receptor multigene family in vertebrates: from the viewpoint of evolutionary genomics. *Curr Genomics.* 2012;13(2):103–114.
12. Nguyen D, Lee K, Choi H, et al. The complete swine olfactory subgenome: expansion of the olfactory gene repertoire in the pig genome. *BMC Genomics.* 2012;13(1):584.
13. Niimura Y, Matsui A, Touhara K. Extreme expansion of the olfactory receptor gene repertoire in African elephants and evolutionary dynamics of orthologous gene groups in 13 placental mammals. *Genome Res.* 2014;24(9):1485–1496.
14. Go Y, Niimura Y. Similar numbers but different repertoires of olfactory receptor genes in humans and chimpanzees. *Mol Biol Evol.* 2008;25(9):1897–1907.
15. Matsui A, Go Y, Niimura Y. Degeneration of olfactory receptor gene repertories in primates: no direct link to full trichromatic vision. *Mol Biol Evol.* 2010;27(5): 1192–1200.
16. Quignon P, Giraud M, Rimbault M, et al. The dog and rat olfactory receptor repertoires. *Genome Biol.* 2005;6(10):R83.
17. Rouquier S, Giorgi D. Olfactory receptor gene repertoires in mammals. *Mutat Res.* 2007;6(16):95–102.
18. Shepherd G. The human sense of smell: are we better than we think? *PLoS Biol.* 2004;2(5):e 146.
19. Mombaerts P. Molecular biology of odorant receptors in vertebrates. *Annu Rev Neurosci.* 1999;22:487–509.
20. Niimura Y, Nei M. Evolution of olfactory receptor genes in the human genome. *Proc Natl Acad Sci.* 2003;100:12235–12240.
21. Trask B, Massa H, Brand-Arpon V, et al. Large multi-chromosomal duplications encompass many members of the olfactory receptor gene family in the human genome. *Hum Mol Genet.* 1998;7(13):2007–2020.

22. Issel-Traver L, Rine J. The evolution of mammalian olfactory receptor genes. *Genetics*. 1997;145(1):185–195.
23. Rouquier S, Friedman C, Cc Delettre, et al. A gene recently inactivated in human defines a new olfactory receptor family in mammals. *Hum Mol Genet*. 1998;7(9): 1337–1345.
24. Sharon D, Glusman G, Pilpel Y, et al. Primate evolution of an olfactory receptor cluster: diversification by gene conversion and recent emergence of speudogenes. *Genomics*. 1999;61(1):24–36.
25. Adipietro K, Mainland J, Matsunami H. Functional evolution of mammalian odorant receptors. *PLoS Genet*. 2012;8(7):e1002821.
26. Man O, Gilad Y, Lancet D. Prediction of the odorant binding site of olfactory receptor proteins by human-mouse comparisons. *Protein Sci*. 2004;13(1):240–254.
27. Niimura Y, Nei M. Evolutionary dynamics of olfactory receptor genes in fishes and tetrapods. *Proc Natl Acad Sci*. 2005;102(17):6039–6044.
28. Niimura Y, Nei M. Comparative evolutionary analysis of olfactory receptor gene clusters between humans and mice. *Gene*. 2005;346:13–21.
29. Nei M, Niimura Y, Nozawa M. The evolution of animal chemosensory receptor gene repertoires: roles of chance and necessity. *Nat Rev Genet*. 2008;9(12):951–963.
30. Malnic B, Godfrey P, Buck L. The human olfactory receptor gene family. *Proc Natl Acad Sci*. 2004;101(8):2584–2589.
31. Sosinsky A, Glusman G, Lancet D. The genomic structure of human olfactory receptor genes. *Genomics*. 2000;70(1):49–61.
32. Lane R, Cutforth T, Young J, et al. Genomic analysis of orthologous mouse and human olfactory receptor loci. *Proc Natl Acad Sci*. 2001;98(13):7390–7395.
33. Lane R, Roach J, Lee I, et al. Genomic analysis of the olfactory receptor region of the mouse and human T-cell receptor alpha/delta loci. *Genome Res*. 2002;12(1): 81–87.
34. Ignatieva E, Levitsky V, Yudin N, Moshkin M, Kolchanov N. Genetic basis of olfactory cognition: extremely high level of DNA sequence polymorphism in promoter regions of the human olfactory receptor genes revealed using the 1000 Genomes Project dataset. *Front Psychol*. 2014;5:Article 247.
35. Young J, Luche R, Trask B. Rigorous and thorough bioinformatic analyses of olfactory receptor promoters confirm enrichment of O/E and homeodomain binding sites but reveal no new common motifs. *BMC Genomics*. 2011;12:561.
36. Bulger M, Bender M, van Doorminck J, et al. Comparative structural and functional analysis of the olfactory receptor genes flanking the human and mouse b-globin gene clusters. *Proc Natl Acad Sci*. 2000;97(26):14560–14565.
37. Michaloski J, Galande P, Malnic B. Identification of potential regulatory motifs in odorant receptor genes by analysis of promoter sequences. *Genome Res*. 2006;16: 1091–1098.
38. Plessy C, Pascarella G, Bertin N, et al. Promoter architecture of mouse olfactory receptor genes. *Genome Res*. 2012;22(3):486–497.
39. Hoppe R, Breer H, Strotmann J. Promoter motifs of olfactory receptor genes expressed in distinct topographic patterns. *Genomics*. 2006;87:711–723.
40. Nishizumi H, Kumasaka K, Inoue N, Nakashima A, Sakano H. Deletion of the core-H region in mice abolishes the expression of three proximal odorant receptor genes in cis. *Proc Natl Acad Sci*. 2007;104(50):20067–20072.
41. Fredriksson R, Lagerström MC, Lundin LG, Schiöth HB. The G-protein-coupled receptors in the human genome form five main families. Phylogenetic analysis, paralogon groups, and fingerprints. *Mol Pharmacol*. 2003;63(6):1256–1272.
42. Clowney E, LeGros M, Mosley C, et al. Nuclear aggregation of olfactory receptor genes governs their monogenic expression. *Cell*. 2012;151(4):724–737.

43. Ferreira T, Wilson S, Choi Y, et al. Silencing of odorant receptor genes by G protein beta-gamma signaling ensures the expression of one odorant receptor per olfactory sensory neuron. *Neuron*. 2014;81(4):847–859.
44. Rodriguez I. Singular expression of olfactory receptor genes. *Cell*. 2013;155(2): 274–277.
45. Magklara A, Yen A, Colquitt B, et al. An epigenetic signature for monoallelic olfactory receptor expression. *Cell*. 2011;145(4):555–570.
46. Armelin-Correa L, Gutiyama L, Brandt D, Malnic B. Nuclear compartmentalization of odorant receptor genes. *Proc Natl Acad Sci*. 2014;111(7):2782–2787.
47. Fuss S, Omura M, Mombaerts P. Local and cis effects of the H element on expression of odorant receptor genes in mouse. *Cell*. 2007;130(2):373–384.
48. Serizawa S, Ishii T, Nakatani H, et al. Mutually exclusive expression of odorant receptor transgenes. *Nat Neurosci*. 2000;3:687–693.
49. Sakano H. Neural map formation in the mouse olfactory system. *Neuron*. 2010;67: 530–542.
50. Fleischmann A, Abdus-Saboor I, Sayed A, Shykind B. Functional interrogation of an odorant receptor locus reveals multiple axes of transcriptional regulation. *PLoS Biol*. 2013;11(5):e1001568.
51. Serizawa S, Miyamichi K, Nakatani H, et al. Negative feedback regulation ensures the one receptor-one olfactory neuron rule in mouse. *Science*. 2003;302(5653):2088–2094.
52. Tan L, Zong C, Xie X. Rare event of histone demethylation can initiate singular gene expression of olfactory receptors. *Proc Natl Acad Sci*. 2013;110(52):21148–21152.
53. Rawson N, Eberwine J, Dotson R, Jackson J, Ulrich P, Restrepo D. Expression of mRNA encoding for two different olfactory receptors in a subset of olfactory receptor neurons. *J Neurochem*. 2000;75(1):185–195.
54. Zhang X, Firestein S. The olfactory receptor gene superfamily of the mouse. *Nat Neurosci*. 2002;5(2):124–133.
55. Miyamichi S, Serizawa H, Kimura H, Sakano H. Continuous and overlapping expression domains of odorant receptor genes in the olfactory epithelium determine the dorsal/ventral positioning of glomeruli in the olfactory bulb. *J Neurosci*. 2005;25: 3586–3592.
56. Kang N, Koo J. Olfactory receptors in non-chemosensory tissues. *BMB Rep*. 2012;45(11):612–622.
57. Flegel C, Manteniotis S, Osthold S, Hatt H, Gisselmann G. Expression profile of ectopic olfactory receptors determined by deep sequencing. *PLoS One*. 2013;8(2): e55368.
58. Foster S, Roura E, Thomas W. Extrasensory perception: odorant and taste receptors beyond the nose and mouth. *Pharmacol Ther*. 2014;142:41–61.
59. Dubacq C, Jamet S, Trembleau A. Evidence for developmentally regulated local translation of odorant receptor mRNAs in the axons of olfactory sensory neurons. *J Neurosci*. 2009;29(33):10184–10190.
60. Richard M, Jamet S, Fouquet C, et al. Homotypic and heterotypic adhesion induced by odorant receptors and the beta2-adrenergic receptor. *PLoS One*. 2013;8(12):e80100.
61. Dubacq C, Fouquet C, Trembleau A. Making scent of the presence and local translation of odorant receptor mRNAs in olfactory axons. *Dev Neurobiol*. 2014;74(3): 259–268.
62. Pronin A, Levay K, Velmeshev D, Faghihi M, Shestopalov V, Slepak V. Expression of olfactory signaling genes in the eye. *PLoS One*. 2014;9(4):e96435.
63. Fukuda N, Touhara K. Developmental expression patterns of testicular olfactory receptor genes during mouse spermatogenesis. *Genes Cells*. 2006;11(1):71–81.
64. Spehr M, Gisselmann G, Poplawski A, et al. Identification of a testicular odorant receptor mediating human sperm chemotaxis. *Science*. 2003;299:2054–2058.

65. Spehr M, Schwane K, Heilmann S, Gisselmann G, Hummel T, Hatt H. Dual capacity of a human olfactory receptor. *Curr Biol.* 2004;14(19):R832–R833.
66. Gakamsky A, Armon L, Eisenbach M. Behavioral response of human spermatozoa to a concentration jump of chemoattractants or intracellular cyclic nucleotides. *Hum Reprod.* 2009;24(5):1152–1163.
67. Griffin CA, Kafadar KA, Pavlath GK. MOR23 promotes muscle regeneration and regulates cell adhesion and migration. *Dev Cell.* 2009;17:649–661.
68. Pluznick JL, Zou DJ, Zhang X, et al. Functional expression of the olfactory signaling system in the kidney. *Proc Natl Acad Sci USA.* 2009;106(6):2059–2064.
69. Pluznick JL, Protzko RJ, Gevorgyan H, et al. Olfactory receptor responding to gut microbiota-derived signals plays a role in renin secretion and blood pressure regulation. *Proc Natl Acad Sci USA.* 2013;110(11):4410–4415.
70. Braun T, Voland P, Kunz L, Prinz C, Gratzl M. Enterochromaffin cells of the human gut: sensors for spices and odorants. *Gastroenterology.* 2007;132(5):1890–1901.
71. Kidd M, Modlin IM, Gustafsson BI, Drozdov I, Hauso O, Pfragner R. Luminal regulation of normal and neoplastic human EC cell serotonin release is mediated by bile salts, amines, tastants, and olfactants. *Am J Physiol Gastrointest Liver Physiol.* 2008;295(2): G260–G272.
72. Primeaux S, Braymer H, Bray G. High fat diet differentially regulates the expression of olfactory receptors in the duodenum of obesity-prone and obesity-resistant rats. *Dig Dis Sci.* 2013;58(1):72–76.
73. Itakura S, Ohno K, Ueki T, Sato K, Kanayama N. Expression of Golf in the rat placenta: possible implication of olfactory receptor transduction. *Placenta.* 2006;27(1):103–108.
74. Cui T, Tsolakis A, Li S, et al. Olfactory receptor 51E1 protein as a potential novel tissue biomarker for small intestine neuroendocrine carcinomas. *Eur J Endocrinol.* 2013;168(2):253–261.
75. Weng J, Wang J, Hu X, Wang F, Ittman M, Liu M. PSGR2, a novel G-protein coupled receptor, is overexpressed in human prostate cancer. *Int J Cancer.* 2006;118(6): 1471–1480.
76. Leja J, Essaghir A, Essand M, et al. Novel markers for enterochromaffin cells and gastrointestinal neuroendocrine carcinomas. *Mod Pathol.* 2009;22:261–272.
77. Muranen TA, Greco D, Fagerholm R, et al. Breast tumors from CHEK2 1100delC-mutation carriers: genomic landscape and clinical implications. *Breast Cancer Res: BCR.* 2011;13(5):R90.
78. Sanz G, Leray I, Dewaele A, et al. Promotion of cancer cell invasiveness and metastasis emergence caused by olfactory receptor stimulation. *PLoS One.* 2014;9(1):e85100.
79. Zhang X, Bedigian AV, Wang W, Eggert US. G protein-coupled receptors participate in cytokinesis. *Cytoskeleton (Hoboken).* 2012;69(10):810–818.
80. Li J, Tay H, Plank M, et al. Activation of olfactory receptors on mouse pulmonary macrophages promotes monocyte chemotactic protein-1 production. *PLoS One.* 2013;8(11):e80148.
81. Saito H, Chi Q, Zhuang H, Matsunami H, Mainland J. Odor coding by a mammalian receptor repertoire. *Sci Signal.* 2009;2(60):ra9.
82. Mainland J, Keller A, Li Y, et al. The missense of smell: functional variability in the human odorant receptor repertoire. *Nat Neurosci.* 2014;17(1):114–120.
83. Peterlin Z, Firestein S, Rogers M. The state of the art of odorant receptor deorphanization: a report from the orphanage. *J Gen Physiol.* 2014;143(5):527–542.
84. Bush C, Jones S, Lyle A, Minneman K, Ressler K, Hall R. Specificity of olfactory receptor interaction with other G protein-coupled receptors. *J Biol Chem.* 2007;282(26):19042–19051.

85. Hague C, Uberti M, Chen Z, et al. Olfactory receptor surface expression is driven by association with the beta2-adrenergic receptor. *Proc Natl Acad Sci.* 2004;101(37): 13672–13676.
86. Saito H, Kubota M, Roberts R, Chi Q, Matsunami H. RTP family members induce functional expression of mammalian odorant receptors. *Cell.* 2004;119(5):679–691.
87. Kaupp U. Olfactory signalling in vertebrates and insects: differences and commonalities. *Nat Rev Neurosci.* 2010;11(3):188–200.
88. Kleene S. The electrochemical basis of odor transduction in vertebrate olfactory cilia. *Chem Senses.* 2008;33(9):839–859.
89. Schild D, Restrepo D. Transduction mechanisms in vertebrate olfactory receptor cells. *Physiol Rev.* 1998;78(2):429–465.
90. Brunet L, Gold G, Ngai J. General anosmia caused by a targeted disruption of the mouse olfactory cyclic nucleotide-gated cation channel. *Neuron.* 1996;17(4):681–693.
91. Belluscio L, Gold G, Nemes A, Axel R. Mice deficient in G(olf) are anosmic. *Neuron.* 1998;20(1):69–81.
92. Jones D, Reed R. Golf: an olfactory neuron specific-G protein involved in odorant signal transduction. *Science.* 1989;244(4906):790–795.
93. Li F, Ponissery-Saidu S, Yee K, et al. Heterotrimeric G protein subunit G gamma13 is critical to olfaction. *J Neurosci.* 2013;33(18):7975–7984.
94. Nakamura T, Gold G. A cyclic nucleotide-gated conductance in olfactory receptor cilia. *Nature.* 1987;325(6103):442–444.
95. Yu Y, Boyer NP, Zhang C. Three structurally similar odorants trigger distinct signaling pathways in a mouse olfactory neuron. *Neuroscience.* 2014;275:194–210.
96. Liu G, Badeau RM, Tanimura A, Talamo BR. Odorant receptors directly activate phospholipase C/inositol-1,4,5-trisphosphate coupled to calcium influx in Odora cells. *J Neurochem.* 2006;96(6):1591–1605.
97. Kaur R, Zhu X, Moorhouse A, Barry P. IP_3-gated channels and their occurrence relative to CNG channels in the soma and dendritic knob of rat olfactory receptor neurons. *J Membr Biol.* 2001;181:91–105.
98. Boekhoff I, Tareilus E, Strotmann J, Breer H. Rapid activation of alternative second messenger pathways on olfactory cilia from rats by different odorants. *EMBO J.* 1990;9(8):2453–2458.
99. Lischka F, Zviman M, Teeter J, Restrepo D. Characterization of inositol-1,4,5-triphosphate-gated channels in the plasma membrane of rat olfactory neurons. *Biophys J.* 1999;76(3):1410–1422.
100. Vogl A, Noé J, Breer H, Boekhoff I. Cross-talk between olfactory second messenger pathways. *Eur J Biochem.* 2000;267(14):4529–4535.
101. Huque T, Bruch R. Odorant- and guanine nucleotide-stimulated phosphoinositide turnover in olfactory cilia. *Biochem Biophys Res Commun.* 1986;137(1):36–42.
102. Restrepo D, Miyamoto T, Bryant B, Teeter J. Odor stimuli trigger influx of calcium in olfactory neurons of the channel catfish. *Science.* 1990;249(4973):1166–1168.
103. Kashiwayanagi M. Dialysis of inositol 1,4,5-triphosphate induces inward currents and Ca2+ uptake in frog olfactory receptor cells. *Biochem Biophys Res Commun.* 1996;225(2):666–671.
104. Hatt H, Ache B. Cyclic nucleotide- and inositol phosphate-gated ion channels in lobster olfactory receptor neurons. *Proc Natl Acad Sci.* 1994;91(14):6264–6268.
105. Delay RJ, Dionne VE. Two second messengers mediate amino acid responses in olfactory sensory neurons of the salamander, Necturus maculosus. *Chem Senses.* 2002;27:673–680.
106. Ache BW. Odorant-specific modes of signaling in mammalian olfaction. *Chem Senses.* 2010;35(7):533–539.

107. Rospars J, Lansky P, Chaput M, Duchamp-Viret P. Competitive and non competitive odorant interactions in the early neural coding of odorant mixtures. *J Neurosci.* 2008;28(10):2659–2666.
108. Ukhanov K, Brunert D, Corey EA, Ache BW. Phosphoinositide 3-kinase-dependent antagonism in Mammalian olfactory receptor neurons. *J Neurosci.* 2011;31(1):273–280.
109. Ukhanov K, Corey EA, Ache BW. Phosphoinositide 3-kinase dependent inhibition as a broad basis for opponent coding in Mammalian olfactory receptor neurons. *PLoS One.* 2013;8(4):e61553.
110. Minic J, Persuy MA, Godel E, et al. Functional expression of olfactory receptors in yeast and development of a bioassay for odorant screening. *FEBS J.* 2005;272(2):524–537.
111. Minic Vidic J, Grosclaude J, Persuy MA, Aioun J, Salesse R, Pajot-Augy E. Quantitative assessment of olfactory receptors activity in immobilized nanosomes: a novel concept for bioelectronic nose. *Lab Chip.* 2006;6:1026–1032.
112. Duchamp-Viret P, Duchamp A, Sicard G. Olfactory discrimination over a wide concentration range. Comparison of receptor cell and bulb neuron abilities. *Brain Res.* 1990;517(1–2):256–262.
113. Grosmaitre X, Vassalli A, Mombaerts P, Shepherd GM, Ma M. Odorant responses of olfactory sensory neurons expressing the odorant receptor MOR23: a patch clamp analysis in gene-targeted mice. *Proc Natl Acad Sci USA.* 2006;103(6):1970–1975.
114. Vidic J, Grosclaude J, Monnerie R, et al. On a chip demonstration of a functional role for Odorant Binding Protein in the preservation of olfactory receptor activity at high odorant concentration. *Lab Chip.* 2008;8(5):678–688.
115. Wade F, Espagne A, Persuy MA, et al. Relationship between homo-oligomerization of a mammalian olfactory receptor and its activation state demonstrated by bioluminescence resonance energy transfer. *J Biol Chem.* 2011;286(17):15252–15259.
116. Matarazzo V, Zsürger N, Guillemot JC, et al. Porcine odorant-binding protein selectively binds to a human olfactory receptor. *Chem Senses.* 2002;27:691–701.
117. Franco R, Casado V, Mallol J, et al. The two-state dimer receptor model: a general model for receptor dimers. *Mol Pharmacol.* 2006;69(6):1905–1912.
118. Gu X, Karp PH, Brody SL, et al. Chemosensory functions for pulmonary neuroendocrine cells. *Am J Respir Cell Mol Biol.* 2013;50(3):637–646.
119. Weng J, Wang J, Cai Y, et al. Increased expression of prostate-specific G-protein-coupled receptor in human prostate intraepithelial neoplasia and prostate cancers. *Int J Cancer.* 2005;113(5):811–818.
120. Fredriksson R, Schioth HB. The repertoire of G-protein-coupled receptors in fully sequenced genomes. *Mol Pharmacol.* 2005;67(5):1414–1425.
121. Lounnas V, Ritschel T, Kelder J, McGuire R, Bywater RP, Foloppe N. Current progress in structure-based rational drug design marks a new mindset in drug discovery. *Comput Struct Biotechnol J.* 2013;5:e201302011.
122. Sanz G, Thomas-Danguin T, Hamdani EH, et al. Relationships between molecular structure and perceived odor quality of ligands for a human olfactory receptor. *Chem Senses.* 2008;33(7):639–653.
123. Moitessier N, Englebienne P, Lee D, Lawandi J, Corbeil CR. Towards the development of universal, fast and highly accurate docking/scoring methods: a long way to go. *Br J Pharmacol.* 2008;153(S1):S7–S26.
124. Palczewski K, Kumasaka T, Hori T, et al. Crystal structure of rhodopsin: a G protein-coupled receptor. *Science.* 2000;289(5480):739–745.
125. Okada T, Sugihara M, Bondar AN, Elstner M, Entel P, Buss V. The retinal conformation and its environment in rhodopsin in light of a new 2.2 A crystal structure. *J Mol Biol.* 2004;342(2):571–583.
126. Choe HW, Kim YJ, Park JH, et al. Crystal structure of metarhodopsin II. *Nature.* 2011;471(7340):651–655.

127. Makino CL, Riley CK, Looney J, Crouch RK, Okada T. Binding of more than one retinoid to visual opsins. *Biophys J.* 2010;99(7):2366–2373.
128. Murakami M, Kouyama T. Crystal structure of squid rhodopsin. *Nature.* 2008;453(7193):363–367.
129. Rasmussen SG, Choi HJ, Rosenbaum DM, et al. Crystal structure of the human beta2 adrenergic G-protein-coupled receptor. *Nature.* 2007;450(7168):383–387.
130. Hanson MA, Cherezov V, Griffith MT, et al. A specific cholesterol binding site is established by the 2.8 A structure of the human beta2-adrenergic receptor. *Structure.* 2008;16(6):897–905.
131. Wacker D, Fenalti G, Brown MA, et al. Conserved binding mode of human beta2 adrenergic receptor inverse agonists and antagonist revealed by X-ray crystallography. *J Am Chem Soc.* 2010;132(33):11443–11445.
132. Rasmussen SG, Choi HJ, Fung JJ, et al. Structure of a nanobody-stabilized active state of the beta(2) adrenoceptor. *Nature.* 2011;469(7329):175–180.
133. Rosenbaum DM, Zhang C, Lyons JA, et al. Structure and function of an irreversible agonist-beta(2) adrenoceptor complex. *Nature.* 2011;469(7329):236–240.
134. Rasmussen SG, DeVree BT, Zou Y, et al. Crystal structure of the beta2 adrenergic receptor-Gs protein complex. *Nature.* 2011;477(7366):549–555.
135. Ring AM, Manglik A, Kruse AC, et al. Adrenaline-activated structure of [bgr]2-adrenoceptor stabilized by an engineered nanobody. *Nature.* 2013;502(7472):575–579.
136. Warne T, Serrano-Vega MJ, Baker JG, et al. Structure of a beta1-adrenergic G-protein-coupled receptor. *Nature.* 2008;454(7203):486–491.
137. Warne T, Moukhametzianov R, Baker JG, et al. The structural basis for agonist and partial agonist action on a beta(1)-adrenergic receptor. *Nature.* 2011;469(7329):241–244.
138. Moukhametzianov R, Warne T, Edwards PC, et al. Two distinct conformations of helix 6 observed in antagonist-bound structures of a beta1-adrenergic receptor. *Proc Natl Acad Sci USA.* 2011;108(20):8228–8232.
139. Warne T, Edwards PC, Leslie AG, Tate CG. Crystal structures of a stabilized beta1-adrenoceptor bound to the biased agonists bucindolol and carvedilol. *Structure.* 2012;20(5):841–849.
140. Christopher JA, Brown J, Dore AS, et al. Biophysical fragment screening of the beta-adrenergic receptor: identification of high affinity arylpiperazine leads using structure-based drug design. *J Med Chem.* 2013;56(9):3446–3455.
141. Jaakola VP, Griffith MT, Hanson MA, et al. The 2.6 angstrom crystal structure of a human A2A adenosine receptor bound to an antagonist. *Science.* 2008;322(5905):1211–1217.
142. Xu F, Wu H, Katritch V, et al. Structure of an agonist-bound human A2A adenosine receptor. *Science.* 2011;332(6027):322–327.
143. Lebon G, Warne T, Edwards PC, et al. Agonist-bound adenosine A2A receptor structures reveal common features of GPCR activation. *Nature.* 2011;474(7352):521–525.
144. Dore AS, Robertson N, Errey JC, et al. Structure of the adenosine A(2A) receptor in complex with ZM241385 and the xanthines XAC and caffeine. *Structure.* 2011;19(9):1283–1293.
145. Wu B, Chien EY, Mol CD, et al. Structures of the CXCR4 chemokine GPCR with small-molecule and cyclic peptide antagonists. *Science.* 2010;330(6007):1066–1071.
146. Chien EY, Liu W, Zhao Q, et al. Structure of the human dopamine D3 receptor in complex with a D2/D3 selective antagonist. *Science.* 2010;330(6007):1091–1095.
147. Shimamura T, Shiroishi M, Weyand S, et al. Structure of the human histamine H1 receptor complex with doxepin. *Nature.* 2011;475(7354):65–70.
148. Hanson MA, Roth CB, Jo E, et al. Crystal structure of a lipid G protein-coupled receptor. *Science.* 2012;335(6070):851–855.

149. Haga K, Kruse AC, Asada H, et al. Structure of the human M2 muscarinic acetylcholine receptor bound to an antagonist. *Nature*. 2012;482(7386):547–551.
150. Kruse AC, Ring AM, Manglik A, et al. Activation and allosteric modulation of a muscarinic acetylcholine receptor. *Nature*. 2013;504(7478):101–106.
151. Kruse AC, Hu J, Pan AC, et al. Structure and dynamics of the M3 muscarinic acetylcholine receptor. *Nature*. 2012;482(7386):552–556.
152. Manglik A, Kruse AC, Kobilka TS, et al. Crystal structure of the micro-opioid receptor bound to a morphinan antagonist. *Nature*. 2012;485(7398):321–326.
153. Wu H, Wacker D, Mileni M, et al. Structure of the human kappa-opioid receptor in complex with JDTic. *Nature*. 2012;485(7398):327–332.
154. Granier S, Manglik A, Kruse AC, et al. Structure of the delta-opioid receptor bound to naltrindole. *Nature*. 2012;485(7398):400–404.
155. Thompson AA, Liu W, Chun E, et al. Structure of the nociceptin/orphanin FQ receptor in complex with a peptide mimetic. *Nature*. 2012;485(7398):395–399.
156. White JF, Noinaj N, Shibata Y, et al. Structure of the agonist-bound neurotensin receptor. *Nature*. 2012;490(7421):508–513.
157. Zhang C, Srinivasan Y, Arlow DH, et al. High-resolution crystal structure of human protease-activated receptor 1. *Nature*. 2012;492(7429):387–392.
158. Park SH, Das BB, Casagrande F, et al. Structure of the chemokine receptor CXCR1 in phospholipid bilayers. *Nature*. 2012;491(7426):779–783.
159. Wacker D, Wang C, Katritch V, et al. Structural features for functional selectivity at serotonin receptors. *Science*. 2013;340(6132):615–619.
160. Tan Q, Zhu Y, Li J, et al. Structure of the CCR5 chemokine receptor–HIV entry inhibitor maraviroc complex. *Science*. 2013;341(6152):1387–1390.
161. Zhang J, Zhang K, Gao Z-G, et al. Agonist-bound structure of the human P2Y12 receptor. *Nature*. 2014;509(7498):119–122.
162. Zhang K, Zhang J, Gao Z-G, et al. Structure of the human P2Y12 receptor in complex with an antithrombotic drug. *Nature*. 2014;509(7498):115–118.
163. Srivastava A, Yano J, Hirozane Y, et al. High-resolution structure of the human GPR40 receptor bound to allosteric agonist TAK-875. *Nature*. 2014;513(7516): 124–127.
164. Nygaard R, Zou Y, Dror RO, et al. The dynamic process of beta(2)-adrenergic receptor activation. *Cell*. 2013;152(3):532–542.
165. Standfuss J, Edwards PC, D'Antona A, et al. The structural basis of agonist-induced activation in constitutively active rhodopsin. *Nature*. 2011;471(7340):656–660.
166. Liu JJ, Horst R, Katritch V, Stevens RC, Wuthrich K. Biased signaling pathways in beta2-adrenergic receptor characterized by 19F-NMR. *Science*. 2012;335(6072): 1106–1110.
167. Venkatakrishnan AJ, Deupi X, Lebon G, Tate CG, Schertler GF, Babu MM. Molecular signatures of G-protein-coupled receptors. *Nature*. 2013;494(7436):185–194.
168. Deupi X, Standfuss J. Structural insights into agonist-induced activation of G-protein-coupled receptors. *Curr Opin Struct Biol*. 2011;21(4):541–551.
169. Hollenstein K, Kean J, Bortolato A, et al. Structure of class B GPCR corticotropin-releasing factor receptor 1. *Nature*. 2013;499(7459):438–443.
170. Wang C, Wu H, Katritch V, et al. Structure of the human smoothened receptor bound to an antitumour agent. *Nature*. 2013;497(7449):338–343.
171. Dore AS, Okrasa K, Patel JC, et al. Structure of class C GPCR metabotropic glutamate receptor 5 transmembrane domain. *Nature*. 2014;511(7511):557–562.
172. Katritch V, Cherezov V, Stevens RC. Structure-function of the G protein-coupled receptor superfamily. *Annu Rev Pharmacol Toxicol*. 2013;53(1):531–556.
173. Kontoyianni M, Liu Z. Structure-based design in the GPCR target space. *Curr Med Chem*. 2012;19(4):544–556.

174. Costanzi S. Modeling G, protein-coupled receptors and their interactions with ligands. *Curr Opin Struct Biol.* 2013;23(2):185–190.
175. Koolstra AJ, Roumen L, Leurs R, de Esch IJ, de Graaf C. From heptahelical bundle to hits from the Haystack: structure-based virtual screening for GPCR ligands. *Methods Enzymol.* 2013;522:279–336.
176. Tautermann CS. GPCR structures in drug design, emerging opportunities with new structures. *Bioorg Med Chem Lett.* 2014;24(17):4073–4079.
177. Xiang Z. Advances in homology protein structure modeling. *Curr Protein Pept Sci.* 2006;7(3):217–227.
178. Launay G, Sanz G, Pajot E, Gibrat JF. Modeling of mammalian olfactory receptors and docking of odorants. *Biophys Rev.* 2012;4:255–269.
179. Ballesteros JA, Weinstein H. Integrated methods and computational probing of structure-function relations in G protein-coupled receptors. *Methods Neurosci.* 1995;25:366–428.
180. Michalsky E, Goede A, Preissner R. Loops In Proteins (LIP)—a comprehensive loop database for homology modelling. *Protein Eng.* 2003;16(12):979–985.
181. Congreve M, Langmead CJ, Mason JS, Marshall FH. Progress in structure based drug design for G protein-coupled receptors. *J Med Chem.* 2011;54(13):4283–4311.
182. Krieger E, Joo K, Lee J, et al. Improving physical realism, stereochemistry, and side-chain accuracy in homology modeling: Four approaches that performed well in CASP8. *Proteins: Struct, Funct, Bioinform.* 2009;77(S9):114–122.
183. Zhukov A, Andrews SP, Errey JC, et al. Biophysical mapping of the adenosine A2A receptor. *J Med Chem.* 2011;54(13):4312–4323.
184. Stouch T. The errors of our ways: taking account of error in computer-aided drug design to build confidence intervals for our next 25 years. *J Comput Aid Mol Des.* 2012;26(1):125–134.
185. Beuming T, Sherman W. Current assessment of docking into GPCR crystal structures and homology models: successes, challenges, and guidelines. *J Chem Inf Model.* 2012;52(12):3263–3277.
186. Kufareva I, Rueda M, Katritch V, Stevens RC, Abagyan R. Status of GPCR modeling and docking as reflected by community-wide GPCR Dock 2010 assessment. *Structure.* 2011;19(8):1108–1126.
187. Shoichet BK, Kobilka BK. Structure-based drug screening for G-protein-coupled receptors. *Trends Pharmacol Sci.* 2012;33(5):268–272.
188. Wichard JD, Ter Laak A, Krause G, Heinrich N, Kühne R, Kleinau G. Chemogenomic analysis of G-protein coupled receptors and their ligands deciphers locks and keys governing diverse aspects of signalling. *PLoS One.* 2011;6(2):e16811.
189. Kenakin T. Functional selectivity and biased receptor signaling. *J Pharmacol Exp Ther.* 2011;336(2):296–302.
190. Kohlhoff KJ, Shukla D, Lawrenz M, et al. Cloud-based simulations on Google Exacycle reveal ligand-modulation of GPCR activation pathways. *Nat Chem.* 2014;6(1):15–21.
191. Launay G, Teletchea S, Wade F, Pajot-Augy E, Gibrat JF, Sanz G. Automatic modeling of mammalian olfactory receptors and docking of odorants. *Protein Eng Des Selection: PEDS.* 2012;25(8):377–386.
192. Selassie CD. History of quantitative structure-activity relationships. In: Abraham DJ, ed. *Burger's medicinal chemistry and drug discovery.* New-York: John Wiley&Sons, Inc; 2003:6th ed. Drug Discovery. Vol. 1.
193. Pauling L. Molecular architecture and biological reactions. *Chem Eng News.* 1946;24(10):1365–1377.
194. Amoore JE. The stereochemical specificities of human olfactory receptors. *Perfum Essent Oil Rec.* 1952;43:321–330.
195. Amoore JE. Stereochemical theory of olfaction. *Nature.* 1963;198:271–272.

196. Hansch C, Fujita T. *p*-σ-Π Analysis. A method for the correlation of biological activity and chemical structure. *J Am Chem Soc.* 1964;86(8):1616–1626.
197. Rossiter KJ. Structure-odor relationships. *Chem Rev.* 1996;96(8):3201–3240.
198. Chastrette M. Trends in structure-odor relationships. *SAR QSAR Environ Res.* 1997;6(3–4):215–254.
199. Teixeira MA, Barrault L, Rodriguez O, Carvalho CC, Rodrigues AE. Perfumery radar 2.0: a step toward fragrance design and classification. *Ind Eng Chem Res.* 2014;53(21):8890–8912.
200. Sell CS. On the unpredictability of odor. *Angew Chem Int Ed.* 2006;45(38):6254–6261.
201. Hau KM, Connell DW. Quantitative structure-activity relationships (QSARs) for odor thresholds of volatile organic compounds (VOCs). *Indoor Air.* Mar 1998;8(1):23–33.
202. Kirsch F, Buettner A. Odor qualities and thresholds of physiological metabolites of 1,8-cineole as an example for structure-activity relationships considering chirality aspects. *Chem Biodivers.* 2013;10(9):1683–1695.
203. Pal P, Mitra I, Roy K. Predictive QSPR modelling for the olfactory threshold of a series of pyrazine derivatives. *Flavour Frag J.* 2013;28(2):102–117.
204. Rodriguez O, Teixeira MA, Rodrigues AE. Prediction of odour detection thresholds using partition coefficients. *Flavour Frag J.* 2011;26(6):421–428.
205. Tan YX, Siebert KJ. Quantitative structure-activity relationship modeling of alcohol, ester, aldehyde, and ketone flavor thresholds in beer from molecular features. *J Agric Food Chem.* 2004;52(10):3057–3064.
206. Tsantili-Kakoulidou A, Kier LB. A quantitative structure activity relationship (QSAR) study of alkylpyrazine odor modalities. *Pharm Res.* 1992;9(10):1321–1323.
207. Buchbauer G, Klein CT, Wailzer B, Wolschann P. Threshold-based structure-activity relationships of pyrazines with bell-pepper flavor. *J Agric Food Chem.* 2000;48(9): 4273–4278.
208. Chastrette M, Cretin D, ElAidi C. Structure-odor relationships: using neural networks in the estimation of camphoraceous or fruity odors and olfactory thresholds of aliphatic alcohols. *J Chem Inf Comput Sci.* 1996;36(1):108–113.
209. Egolf LM, Jurs PC. Quantitative structure-retention and structure-odor intensity relationships for a diverse group of odor-active compounds. *Anal Chem.* 1993;65(21):3119–3126.
210. Pal P, Mitra I, Roy K. A quantitative structure–property relationship approach to determine the essential molecular functionalities of potent odorants. *Flavour Frag J.* 2014;29(3):157–165.
211. Wailzer B, Klocker J, Buchbauer G, Ecker G, Wolschann P. Prediction of the aroma quality and the threshold values of some pyrazines using artificial neural networks. *J Med Chem.* 2001;44(17):2805–2813.
212. Zakarya D, Farhaoui L, Hamidi M, Bouachrine M. Structure-olfactive threshold relationships for pyrazine derivatives. *J Mol Model.* 2006;12(6):985–989.
213. Haddad R, Khan R, Takahashi YK, Mori K, Harel D, Sobel N. A metric for odorant comparison. *Nat Methods.* 2008;5(5):425–429.
214. Kaeppler K, Mueller F. Odor classification: a review of factors influencing perception-based odor arrangements. *Chem Senses.* 2013;38(3):189–209.
215. Mahlke IT, Thiesen PH, Niemeyer B. Chemical indices and methods of multivariate statistics as a tool for odor classification. *Environ Sci Technol.* 2007;41(7):2414–2421.
216. Martinez-Mayorga K, Peppard TL, Yongye AB, Santos R, Giulianotti M, Medina-Franco JL. Characterization of a comprehensive flavor database. *J Chemometr.* 2011;25(10):550–560.
217. Omatu S, Araki H, Fujinaka T, Yoshioka M, Nakazumi H. Mixed odor classification for QCM sensor data by neural networks. In: Omatu S, Santana JFD, Gonzalez SR, Molina JM, Bernardos AM, Rodriguez JMC, eds. *Distributed Computing and Artificial Intelligence.* Berlin, Germany: Springer-Verlag; 2012:1–8; Vol. 151.

218. Ros F, Audouze K, Pintore M, Chretien JR. Hybrid systems for virtual screening: interest of fuzzy clustering applied to olfaction. *SAR QSAR Environ Res.* 2000;11(3–4):281–300.
219. Snitz K, Yablonka A, Weiss T, Frumin I, Khan RM, Sobel N. Predicting odor perceptual similarity from odor structure. *PLoS Comput Biol.* 2013;9(9):e1003184.
220. Chastrette M, de Saint Laumer J. Structure-odor relationships using neural networks. *Eur J Med Chem.* 1991;26(8):829–833.
221. Chastrette M, Zakarya D, Peyraud J. Structure-musk odor relationships for tetralins and indans using neural networks (on the contribution of descriptors to the classification). *Eur J Med Chem.* 1994;29(5):343–348.
222. Klocker J, Wailzer B, Buchbauer G, Wolschann P. Aroma quality differentiation of pyrazine derivatives using self-organizing molecular field analysis and artificial neural network. *J Agric Food Chem.* 2002;50(14):4069–4075.
223. Audouze K, Ros F, Pintore M, Chretien JR. Prediction of odours of aliphatic alcohols and carbonylated compounds using fuzzy partition and self organising maps (SOM). *Analusis.* 2000;28(7):625–632.
224. Takane SY, Mitchell JBO. A structure-odour relationship study using EVA descriptors and hierarchical clustering. *Org Biomol Chem.* 2004;2(22):3250–3255.
225. Khan RM, Luk CH, Flinker A, et al. Predicting odor pleasantness from odorant structure: pleasantness as a reflection of the physical world. *J Neurosci.* 2007;27(37):10015–10023.
226. Kraft P, Bajgrowicz JA, Denis C, Frater G. Odds and trends: recent developments in the chemistry of odorants. *Angew Chem Int Ed.* 2000;39(17):2981–3010.
227. Lavine BK, Davidson CE, Breneman C, Katt W. Electronic van der Waals surface property descriptors and genetic algorithms for developing structure-activity correlations in olfactory databases. *J Chem Inf Comput Sci.* 2003;43(6):1890–1905.
228. Amboni R, da Silva Junkes B, Yunes RA, Fonseca Heinzen VE. Quantitative structure-odor relationships of aliphatic esters using topological indices. *J Agric Food Chem.* 2000;48(8):3517–3521.
229. Zakarya D, Chastrette M, Tollabi M, Fkih-Tetouani S. Structure-camphor odour relationships using the Generation and Selection of Pertinent Descriptors approach. *Chemometr Intell Lab Syst.* 1999;48(1):35–46.
230. Chastrette M, Rallet E. Structure-minty odour relationships: suggestion of an interaction pattern. *Flavour Frag J.* 1998;13(1):5–18.
231. Chastrette M, AIdi CE, Peyraud J. Tetralin, indan and nitrobenzene compound structure-musk odor relationship using neural networks. *Eur J Med Chem.* 1995;30(9):679–686.
232. Buchbauer G, Stappen I, Pretterklieber C, Wolschann A. Structure-activity relationships of sandalwood odorants: synthesis and odor of tricyclo beta-santalol. *Eur J Med Chem.* 2004;39(12):1039–1046.
233. Cheng LP, Xu L, Mao HF, Wang GL. Study of structural and electronic origin of ambergris odor of some compounds. *J Mol Model.* 2009;15(1):1–8.
234. Cheng L, Gu Y, Jin QA. QSAR of alpha-campholenic derivatives with sandalwood odor, and molecular design. *Mon Chem.* 2010;141(9):953–959.
235. Hadaruga DI, Muresan S, Bologa C, et al. QSAR for cycloaliphatic alcohols with qualitatively defined sandalwood odour characteristics. *Quant Struct-Act Relat.* 1999;18(3):253–261.
236. Kovatcheva A, Buchbauer G, Golbraikh A, Wolschann P. QSAR modeling of alpha-campholenic derivatives with sandalwood odor. *J Chem Inf Comput Sci.* 2003;43(1):259–266.
237. Kovatcheva A, Golbraikh A, Oloff S, Feng J, Zheng W, Tropsha A. QSAR modeling of datasets with enantioselective compounds using chirality sensitive molecular descriptors. *SAR QSAR Environ Res.* 2005;16(1–2):93–102.

238. Shvets NM, Dimoglo AS. Structure-odour relationships: results of an applied electron-topological approach. *Nahrung*. 1998;42(6):364–370.
239. Stappen I, Hofinghoff J, Friedl S, Pammer C, Wolschann P, Buchbauer G. Structure-activity relationships of sandalwood odorants: total synthesis and fragrance properties of cyclopropano-beta-santalol. *Eur J Med Chem*. 2008;43(7):1525–1529.
240. Gorbachov MY, Rossiter KJ. A new electronic-topological investigation of the relationship between chemical structure and ambergris odour. *Chem Senses*. 1999;24(2):171–178.
241. Kovatcheva A, Golbraikh A, Oloff S, et al. Combinatorial QSAR of ambergris fragrance compounds. *J Chem Inf Comput Sci*. 2004;44(2):582–595.
242. Chastrette M, Zakarya D, Pierre C. Relations structure-odeur de bois de santal: recherche d'un modele d'interaction fonde sur le concept d'hypermotif santalophore. *Eur J Med Chem*. 1990;25(5):433–440.
243. Rognon C, Chastrette M. Structure-odor relationships: a highly predictive tridimensional interaction model for the bell-pepper note. *Eur J Med Chem*. 1994;29(7–8):595–609.
244. Leach AR, Gillet VJ, Lewis RA, Taylor R. Three-dimensional pharmacophore methods in drug discovery. *J Med Chem*. 2010;53(2):539–558.
245. Bajgrowicz JA, Frater G. Chiral recognition of sandalwood odorants. *Enantiomer*. 2000;5:225–234.
246. Bajgrowicz JA, Berg-Schultz K, Brunner G. Substituted hepta-1,6-dien-3-ones with green/fruity odours green/galbanum olfactophore model. *Bioorg Med Chem*. 2003;11(13):2931–2946.
247. Furrer SM, Slack JP, McCluskey ST, et al. New developments in the chemistry of cooling compounds. *Chemosens Percept*. 2008;1(2):119–126.
248. Svitanko IV, Devetyarov DA, Tcheboukov DE, et al. QSAR modeling on the basis of 3D descriptors representing the electrostatic molecular surface (ambergris fragrances). *Mendeleev Commun*. 2007;17(2):90–91.
249. Yoshii F, Hirono S. Construction of a quantitative three-dimensional model for odor quality using comparative molecular field analysis (CoMFA). *Chem Senses*. 1996;21(2):201–210.
250. Audouze K, Tromelin A, Le Bon AM, et al. Identification of odorant-receptor interactions by global mapping of the human odorome. *PLoS One*. 2014;9(4):e93037.
251. Medina-Franco JL, Martinez-Mayorga K, Peppard TL, Del Rio A. Chemoinformatic analysis of GRAS (generally recognized as safe) flavor chemicals and natural products. *PLoS One*. 2012;7(11):e50798.
252. Valerio LG, Choudhuri S. Chemoinformatics and chemical genomics: potential utility of in silico methods. *J Appl Toxicol*. 2012;32(11):880–889.
253. Virshup AM, Contreras-Garcia J, Wipf P, Yang WT, Beratan DN. Stochastic voyages into uncharted chemical space produce a representative library of all possible drug-like compounds. *J Am Chem Soc*. 2013;135(19):7296–7303.

CHAPTER TWO

Olfactory Signaling in Insects

Dieter Wicher[1]
Max Planck Institute for Chemical Ecology, Department Evolutionary Neuroethology, Jena, Germany
[1]Corresponding author: e-mail address: dwicher@ice.mpg.de

Contents

Abstract

The detection of volatile chemical information in insects is performed by three types of olfactory receptors, odorant receptors (ORs), specific gustatory receptor (GR) proteins for carbon dioxide perception, and ionotropic receptors (IRs) which are related to ionotropic glutamate receptors. All receptors form heteromeric assemblies; an OR complex is composed of an odor-specific OrX protein and a coreceptor (Orco). ORs and GRs have a 7-transmembrane topology as for G protein-coupled receptors, but they are inversely inserted into the membrane. Ligand-gated ion channels (ionotropic receptors) and ORs operate as IRs activated by volatile chemical cues. ORs are evolutionarily young receptors, and they first appear in winged insects and seem to be evolved to allow an insect to follow sparse odor tracks during flight. In contrast to IRs, the ORs can be sensitized by repeated subthreshold odor stimulation. This process involves metabotropic signaling. Pheromone receptors are especially sensitive and require an accessory protein to detect the lipid-derived pheromone molecules. Signaling cascades involved in pheromone detection depend on intensity and duration of stimuli and underlie a circadian control. Taken together, detection and processing of volatile information in insects involve ionotropic as well as metabotropic mechanisms. Here, I review the cellular signaling events associated with detection of cognate ligands by the different types of odorant receptors.

Progress in Molecular Biology and Translational Science, Volume 130
ISSN 1877-1173
http://dx.doi.org/10.1016/bs.pmbts.2014.11.002

1. INTRODUCTION

Chemoreception creates an internal representation of an external world built by an almost infinite variety of different molecules. Volatile molecules are detected by olfactory receptors, and molecules in solution are perceived by GRs. This review is focused on structure, function, and regulation of insect olfactory receptors. Functional aspects of olfaction-related receptors in invertebrates and mammals have been previously reviewed.[1–3] For reviews on insect chemosensation including gustatory reception, see Refs. 4,5.

Olfactory receptors are expressed by olfactory sensory neurons (OSNs). These sensory cells are localized in appendages of the forehead between the complex eyes, the antennae (Fig. 1A and B), and maxillary palps. An antenna

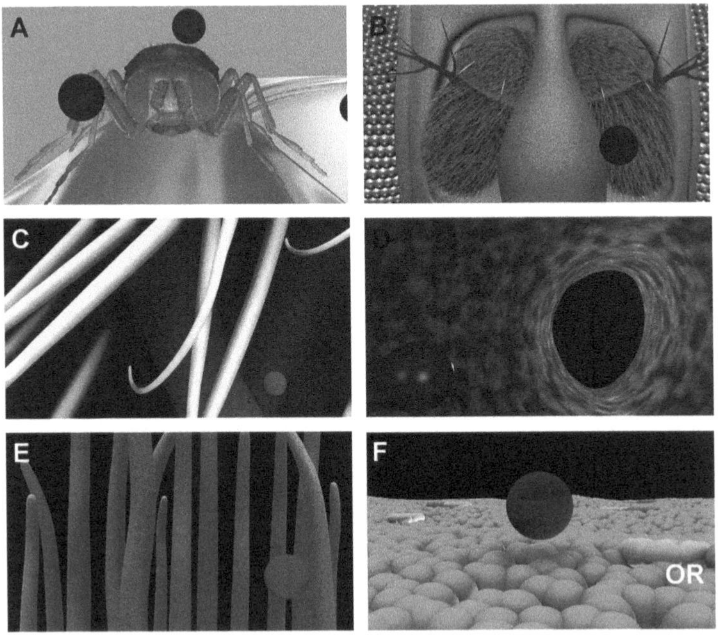

Figure 1 Pathway of odor molecules travelling from the external environment to an olfactory receptor. Odor molecules symbolized by red spheres approach a fly (A) the paired antennae of which (B) are equipped with sensilla (C). The surface of these structures contains pores (C, D) that allow odor molecules to reach the interior of sensilla where the dendrites of olfactory sensory neurons are localized (E). The dendritic membrane contains heterodimeric receptors (ORs) that bind the odorant molecules (F). *Figures are video stills taken from* http://vimeo.com/45493717 *Copyright: © 2012 Kimberly Falk.* (See the color plate.)

is covered by hair-like structures called sensilla (Fig. 1C) that house the dendrites of the OSNs (Fig. 1E). The odor molecules enter a sensillum through pores (Fig. 1D) and bind to their receptors in the dendritic membrane (Fig. 1F). The odor transfer, especially for hydrophobic compounds, is facilitated by specific binding proteins, odorant binding proteins (Fig. 5), or pheromone binding proteins (PBPs; Fig. 6).[6]

The OSNs send their axons into the antennal lobe, the first central nervous processing unit for olfactory information. Similar to the mammalian counterpart, the olfactory bulb, the antennal lobe has a glomerular substructure. The axons of all OSNs expressing a specific receptor target one glomerulus to form synapses with projection neurons that transfer the information into higher brain centers, the mushroom bodies, and the lateral horn. Within the antennal lobe, local interneurons are involved in odor information processing, for example, by linking different glomeruli for combinatorial odor coding.[7] Olfactory signal processing in the central nervous system is reviewed by Ref. 8. For *Drosophila*, there is profound knowledge on odor specificity of receptors.[9] In special cases, receptor stimulation initiates innate behavioral responses such as attraction and aversion.[10] Geosmin, for example, is released by certain toxin-producing bacteria and fungi; for flies, it serves as a danger signal and activates an avoidance reaction.[11] Evolutionary and ecological aspects of insect olfaction are discussed in Refs. 12,13.

The remarkable sensitivity of the insect sense of smell raises the question how this sensory machinery has to be organized to allow an insect such as a moth to detect just a couple of pheromone molecules in a complex environment. One obvious way to enhance the olfactory resolution is to expand the surface of the receptive area, as seen in the large variety of shapes and sizes of insect antennae.[13] The physical limit of the precision in determining the concentration of odor molecules, the Berg–Purcell limit, is determined by the noise due to random diffusion and inversely related to receptor size and sampling time.[14]

A further way to enhance the sensitivity of odor perception is the amplification of the primary chemosignal. During evolution, various changes in the choice of receptor type, ionotropic or metabotropic, and in the downstream amplification mechanism, could be observed.[15] Many sensory systems use G protein-coupled receptors (GPCRs) which provide various levels of amplification. The first-level signaling molecules, heterotrimeric G proteins, can activate enzyme molecules when activated. These enzymes, at the second level, can produce messenger molecules which may induce further downstream signal amplification. Mammalian odorant receptors are GPCRs, as well as receptors for taste and smell in the nematode *Caenorhabditis elegans*.[1] Activated mammalian odorant receptors couple to stimulatory G proteins ($G\alpha_s$) that enhance adenylyl cyclase function and thereby raise the cellular cAMP level (for recent review, see Ref. 3). These second messenger molecules activate cyclic-nucleotide-gated (CNG) channels. CNG channels are nonselective cation channels that conduct Na^+, K^+, and Ca^{2+}. The rise in intracellular Ca^{2+} stimulates Ca^{2+}-activated Cl^- channels which leads to a strong depolarization. Remarkably, the most pronounced amplification of the odor signal within the whole signal transduction cascade is provided by these Cl^- channels.[16]

2. INSECT OLFACTORY RECEPTORS

2.1. Structure

In most insects, volatile signals are perceived by three types of olfactory receptors, odorant receptors (ORs), specific GR proteins for carbon dioxide perception, and so-called ionotropic receptors (IRs) which are related to ionotropic glutamate receptors (iGluRs) (Fig. 2). ORs are heteromers composed of a variable, odor-specific OrX protein or pheromone-specific OrY protein, and a constant coreceptor protein Orco.[17,18] There are three Gr genes for carbon dioxide perception which are absent from basal insects as well as from the European honeybee *Apis mellifera*.[19] These receptors are also heteromers of Gr1/2 and Gr3, in *Drosophila* called Gr21a and Gr63a.[20,21] All OR and GR proteins belong to the 7-transmembrane (7-TM) domain family, but in contrast to GPCR proteins, they show an inversed membrane insertion, i.e., cytoplasmic N-terminus and extracellular C-terminus.[22,23] Finally, the IRs consist of a variable, odor-specific IRX protein and a coreceptor IRcoY protein. The IR coreceptor proteins in *Drosophila* are IR8a and IR25a. Bearing both an extended amino-terminal domain and a ligand-binding domain (LBD), the coreceptor proteins show the most pronounced similarity with iGluRs. As for iGluRs, IRs form heterotetrameric complexes and can contain up to three different subunits. The first IR tetramer formed by dimerization of IR dimers was shown for *Drosophila* IR84a and IR8a.[24]

While IRs and GRs appeared early in evolution and are expressed in a variety of organisms,[25,26] the occurrence of ORs is restricted to insects.[27] Basal nonwinged insects such as bristletails (Orders Thysanura and Archaeognatha) possess no ORs (Archaeognatha) or only Orco-related

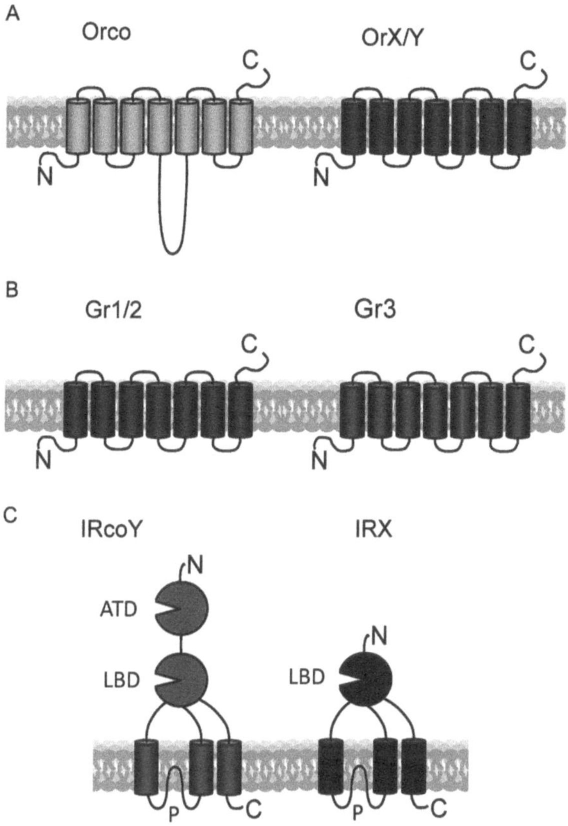

Figure 2 Structure of insect olfactory receptor proteins. (A) Odorant receptors (ORs) are heteromers composed of an ubiquitous coreceptor Orco and an odor-specific Or protein, OrX for food odors (also including odors from oviposition sites, predators, toxic substances, etc.) and OrY for pheromones. These proteins display a 7-transmembrane topology with intracellular N-terminus and extracellular C-terminus, i.e., an opposite orientation compared with GPCRs. (B) Gustatory receptors for carbon dioxide perception form heterodimers of Gr1 or Gr2, and Gr3. The Gr proteins have a structure and orientation similar to Or proteins. (C) Ionotropic receptors are heterotetramers comprising a coreceptor protein IRcoY and a receptor protein IRX. IRs can form tetramers composed of two IRcoY:IRX dimers, but can also contain three different IRX proteins. Both IRcoY and IRX proteins possess three membrane-spanning helices separated by an extracellular region contributing the ligand-binding domain (LBD). P indicates the pore region. The coreceptor proteins have an extended amino-terminal domain (ATD).

proteins (Thysanura). The heteromeric ORs evolved in parallel with the onset of insect flight.[27] This may indicate that the development of ORs is the evolutionary response to the challenge of detecting olfactory information during flight. In addition to the task of tracking odors during motion, odors are dispersed and diluted by the motion of air, thus forming a shifting and filamentous plume. To perceive such information in motion requires fast and sensitive receptive elements.

2.2. Function

As for the related iGluRs, IRs are ligand-gated ion channels. These non-selective cation channels conduct Na^+ and K^+, some also Ca^{2+}.[24] To elicit an odor response, assemblies between IRX and IRcoY are required.[24] As for ORs (see below), the odor specificity is determined by the IRX partner. This may surprise since IRcoY proteins also possess a LBD (Fig. 2). However, as a point mutation in the LBD of IR8a abolished the dendritic localization, the LBD of coreceptors might have another function than binding odorant ligands.[24] Odors may enhance the OSN activity, but they can also reduce it. IRs and ORs detect odors with complementary specificity, i.e., there is less overlap. OSNs expressing IRs are less sensitive than OR-expressing neurons, but IR-expressing neurons adapt more slowly[28] (Fig. 3). For a recent detailed review on insect IRs, see Ref. 29.

In *Drosophila*, the carbon dioxide sensing Gr21a and Gr63a couple to $G\alpha_q$ proteins which activate the phospholipase C (PLC) and in turn ion channels belonging to the transient receptor potential (TRP) family via an as yet unknown mechanism[30,31] (Fig. 4). Acidic odors and high concentrations (>5%) of carbon dioxide are detected by a member of the IR family, IR64a.[32] Carbon dioxide detection is thus an example for a concentration-dependent switch in the primary detection mechanism. Low concentrations are detected via a metabotropic process that provides signal amplification, whereas high concentrations directly activate the ionotropic process.

According to their 7-TM topology, ORs were assumed to operate as GPCR as in nematodes and humans. In line with this, various G proteins are expressed in antennal tissue,[33] and heterologously expressed ORs show functional coupling to G proteins.[34,35] On the other hand, the assumption ORs may act as classical GPCRs was challenged by detecting their inverted membrane insertion.[22,23] Furthermore, there is no sequence similarity between ORs and GPCRs, including the vertebrate ORs.[22] An evolutionary analysis of GPCRs and insect ORs/GRs revealed that the latter families are not closely related to GPCRs.[36]

Odor stimulation of insect ORs expressed in cultured cells demonstrated that they operate as ionotropic receptors.[37,38] Odor binding quickly opens a nonselective cation channel permeable to Na^+, K^+, and Ca^{2+}. Even in the absence of any stimulation, the ORs display constitutive activity leading to

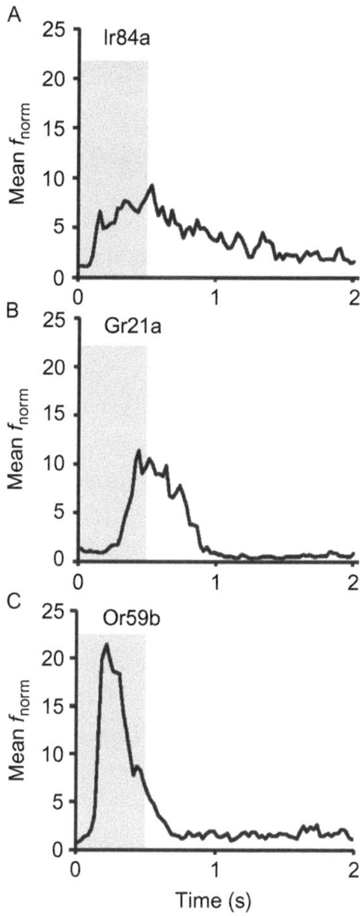

Figure 3 Response of OSNs expressing different olfactory receptor types to odor stimulation given as mean peristimulus histograms. (A) IR84a-expressing neurons stimulated with butyric acid (log [−3] v/v). (B) Gr21a-expressing neurons exposed to pure CO_2. (C) Or59b-expressing OSNs stimulated with methyl acetate (log [−5] v/v). The 500 ms stimuli started at 0 (gray field). The OR-expressing neuron shows the fastest and strongest response to stimulation, but it also displays a fast decay starting during the odor presentation. By contrast, the response of the IR-expressing and the GR-expressing neuron lasts longer and exceeds the time of the stimulus. *Data were taken in modified form from Ref. 28.*

elevated cellular level of free Ca^{2+}. A short odor stimulation was seen to produce an immediate transient response that was independent of G protein activity; the ORs thus act as ligand-gated channels. This type of response requires a sufficient ligand concentration.[37,38] Various models of OR function have been extensively discussed.[39–43] The odorant specificity is

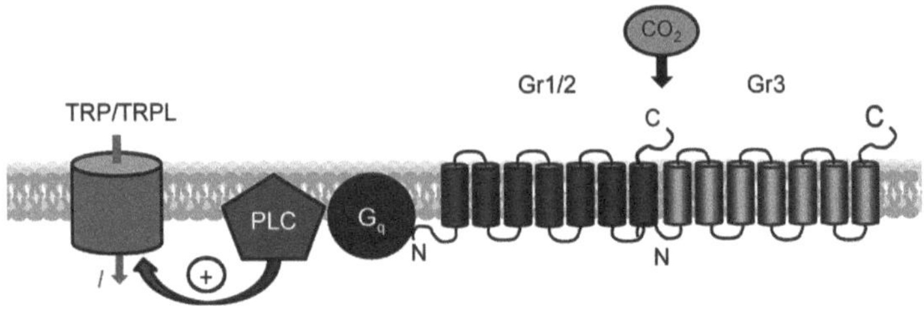

Figure 4 Signal transduction associated with sensing of carbon dioxide by GRs. Binding of CO_2 to the heterodimeric receptor activates a $G\alpha_q$ protein (G_q) which stimulates phospholipase C (PLC). Via an as yet unknown mechanism, PLC activity leads to opening of TRP channels that allow influx of cations (*I*) and thereby depolarize the GR-expressing OSN.

determined by the OrX partner.[9,44] An exchange of only a few amino acids can lead to a shift in agonist specificity as observed for Or22a in *Drosophila melanogaster* versus *Drosophila sechellia*.[45] In this example, the odor preference is shifted from ethyl hexanoate (*D. melanogaster*) to methyl hexanoate (*D. sechellia*).

As there is no crystal structure of ORs available as yet, the topology of these channels remains an open question. They may form a channel as a heterodimer with Orco, similar to the 7-TM protein channelrhodopsin, in which the ion channel pore is formed by oppositely aligned TM3 and TM4 helices.[40,46] Another possible arrangement would be a tetramer composed of two dimers, as, for example, seen in the store-operated Orai channels.[47] The composition of the ORs determines the relative cation permeability, i.e., for different OrX proteins a variable ion permeability was found for the resulting OR complexes.[37,48] Mutational analysis in silkmoth (*Bombyx mori*) ORs revealed that the ion channel pore is lined both by OrX and Orco proteins (Fig. 5).[49] Expression of Orco proteins without OrX proteins also leads to the formation of functional channels which do not respond to odors but can be activated by cyclic nucleotides[38] or synthetic OR agonists.[50–52] Homomeric interactions in cultured cells were in addition to Orco proteins also observed for OrX proteins.[53]

A contribution of intracellular signaling to OR function, including G protein activation, is still controversially discussed. One study employing heterologously expressed ORs and pheromone receptors observed solely ionotropic function.[37] Manipulation of G protein signaling in *Drosophila* had no effect on the OSN activity except for cells expressing the carbon dioxide sensing GRs.[31] Similarly, G protein inhibition in cultured cells expressing OR proteins had little effect on responses to OR stimulation.[54]

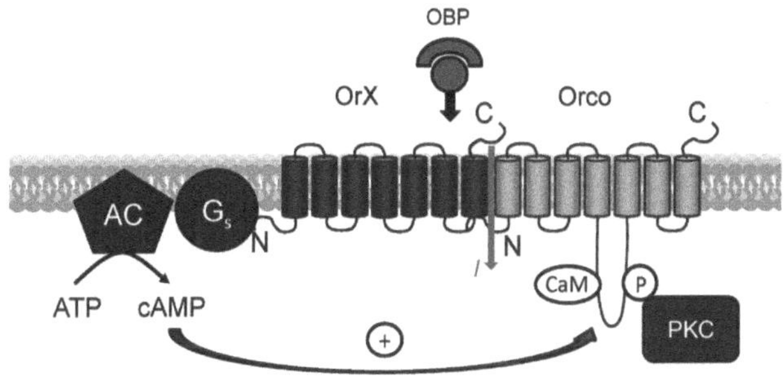

Figure 5 Activation and regulation of food-odor sensing ORs. Odor molecules attached to odor-binding proteins (OBP) approach the OR (OrX), thereby activating an ionotropic current (*I*) upon interaction of ligand and receptor. In parallel, a stimulatory G protein ($G\alpha_s$) is activated which leads to enhanced adenylyl cyclase (AC) activity and thus to increased cAMP production within the cell. Orco channels are activated by cAMP; Orco proteins may form homomeric channels or pass an ion current when paired with a given OrX protein. Orco activity sensitizes ORs in a certain time window after exposure to a ligand. Orco activity is also regulated by phosphorylation via protein kinase C (PKC) and by calmodulin (CaM).

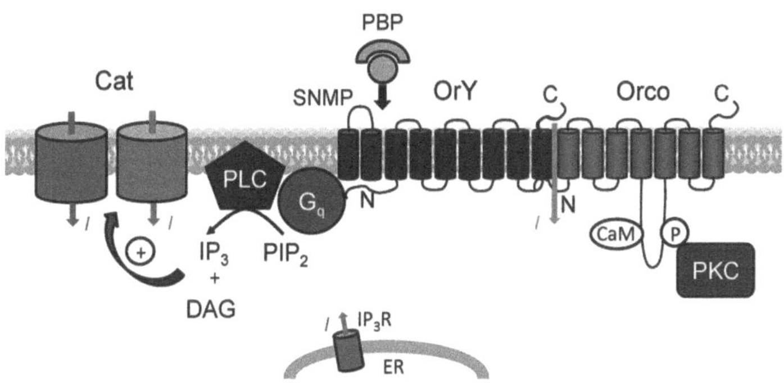

Figure 6 Activation and regulation of pheromone-sensing ORs. The sensory neuron membrane protein (SNMP) acts as coreceptor. Odor molecules attached to pheromone binding proteins (PBPs) approach the OR (OrY), thereby activating an ionotropic current (*I*) upon interaction of ligand and receptor. Alternatively or in parallel, a $G\alpha_q$ protein (G_q) is activated which stimulates phospholipase C (PLC) activity, and thus the production of inositol trisphosphate (IP_3) and diacylglycerol (DAG) from the membrane phospholipid PIP_2. These compounds activate cation channels (Cat) in the plasma membrane. In addition, IP_3 may activate IP_3 receptors (IP_3R) in the endoplasmic reticulum (ER) which causes Ca^{2+} release into the cytoplasm. These events may modulate Orco channel activity via Ca^{2+}/CaM and/or phosphorylation by activation of protein kinase C (PKC) by DAG.

On the other hand, investigation of pheromone reception in the hawkmoth *Manduca sexta* clearly showed metabotropic signaling involving the DAG/IP_3 pathway (Fig. 6).[55] A role of this pathway in mediating olfactory signals was also seen for other insects.[56] The transduction of *Manduca* pheromone signals is controlled by the circadian rhythm and uses various signaling cascades acting on different ion channels.[57] Interestingly, a recent study found no evidence for ionotropic effects in *Manduca* pheromone responses.[58]

One study investigating heterologously expressed *Drosophila* OR proteins found both kinds of signaling, ionotropic as well as metabotropic.[38] In parallel to direct OR channel activation, odor stimulation activated a $G\alpha_s$ protein which enhanced adenylyl cyclase activity and thus the cAMP production.[38] This happens even at odor concentrations too low to activate the ionotropic response. cAMP in turn induces an ion current related to Orco activity. As heterologously expressed Orco proteins form cAMP-activated channels[38] and show homomeric interaction without OrX expression,[53] these Orco channels might be Orco homodimers. In line with this, *Drosophila* Orco-dimer constructs form functional calcium-conducting ion channels when heterologously expressed.[59] These constructs are also modulated by CaM as for Orco proteins and they interact with OrX proteins. An alternative hypothesis for the localization of the Orco channel assumed that the Orco within the OR complex could form another pore[38] in addition to the one facilitating the ionotropic pathway.[37,49]

Given an initial presentation of a subthreshold odor concentration that fails to initiate the ionotropic response but activates the Orco channel, a second presentation of this stimulus, within an appropriate time window between a couple of seconds and a few minutes, will lead to an ionotropic response.[60] A first contact with a weak odor signal thus sensitizes the ORs and qualifies them to respond upon a second, or any further, contact. This time-dependent sensitization represents a short-term memory even at receptor level.

Pheromone perception is a highly sensitive form of olfactory signal processing and of vital importance for successful reproduction in many insects. The hydrophobic nature of pheromones leads to special requirements of the detection apparatus. In addition to pheromone receptors such as Or67d/Orco in *Drosophila*, receptor neurons also express a CD36-related protein, the sensory neuron membrane protein (SNMP).[61] Mammalian CD36 proteins are involved in fatty acid transport and in insect pheromone perception, SNMP is a necessary coreceptor. Prior to binding to the receptor complex, pheromone molecules have to be transferred from the sensillar pores, through the lymph, to the receptor complex. This process is supported by PBPs; for the perception of the *Drosophila* pheromone, *cis*-vaccenyl acetate (cVA), the PBP LUSH acts as carrier.[62] The idea that

LUSH undergoes a conformational change after cVA binding which allows to attach the relevant OrX[62] was not supported by later investigations.[63]

Pheromone receptors of *M. sexta* are known to sense even single pheromone molecules. The pheromone signal processing depends on the signal strength and duration, the circadian rhythm, and the activity level of the insect.[57] Weak stimuli activate the PLC cascade and lead to an increase in cytoplasmic Ca^{2+} concentration (Fig. 6). Strong stimulation activates receptor guanylyl cyclases which causes long-lasting adaptation in OSN activity. In the resting phase, these cells show an enhanced cGMP level, whereas in the activity phase, the cAMP level is high, which leads to differential activation of ion channels. The pheromone signal transduction process is considered as a solely metabotropic one with the pheromone receptor/Orco complex acting as pacemaker channel that drives the OSN membrane potential toward the action potential threshold, thereby accelerating the OSN activity.[57,58] By contrast, heterologously expressed pheromone receptors of *B. mori* were seen to form ligand-gated ion channels, i.e., operated as ionotropic receptors.[37] These divergent observations might reflect the complementary nature of olfactory signaling that involves both, ionotropic as well as metabotropic signaling.

2.3. Regulation

The phenomenon of membrane protein involvement in the transfer of external information into an intracellular signal underlying a regulatory control is well known for ion channels and receptor proteins. A tight regulation of olfactory receptor sensitivity is especially important for insects as they perceive information of vital importance from volatile signal molecules. This includes, for example, signals related to food, partners, enemies, or ovipositioning sites.

A regulation of OR function according to the history as described above is missing for the IRs.[60] Interesting in this context is the correlation between OR evolution and the onset of insect flight. For a proper resolution of filamentous odor plumes during flight, the equipment of OSNs might have to be complemented with the tunable ORs.

The OR sensitization requires previous Orco activity which is controlled by a couple of factors.[60] Orco activation by cAMP, for example, relies on a sufficient degree of Orco phosphorylation via protein kinase C (PKC), which requires a normal physiological intracellular calcium level.[64] On the other hand, highly phosphorylated Orco carries a current even in absence of cAMP. Phosphorylation and cAMP are thus parts of convergent signaling/regulatory pathways.

Mutations in parts of putative odor signal transduction cascades, and observation of the resultant effect on odor responses, provide significant information on determinants of odor perception. *Drosophila* mutants characterized by abnormal cAMP signaling (*dunce* and *rutabaga*) showed disturbed olfactory behavior.[65] Such abnormal behavior was observed when overexpressing the phosphodiesterase *dunce* in OSN subsets which decreased the cAMP level.[66] An important role of $G\alpha_s$ proteins for sensitive OR signaling was also shown.[67] In conclusion, basal function of olfactory signaling relies on an intact cAMP signaling cascade. The processing of olfactory information was also seen to be influenced by other G proteins such as $G\alpha_o$[68] and $G\alpha_i$.[69]

Drosophila norpA mutants display deficits in olfaction which indicates that PLC plays an important role in odor information processing.[70] Mutations in the *dgq* gene which encodes the $G\alpha_q$ subunit produce weaker responses to odor stimulation.[71] These responses are additionally attenuated by mutations in *plc21C*, a gene that encodes for a PLC. Intriguingly, the *dgq* mutant phenotype was rescued by mutation of the DAG kinase *rdgA*[71] which may indicate that the efficiency of the olfactory signal transduction could be controlled by a phospholipid messenger. This hypothesis comes from a study on mutants in *stmbhA*, a gene encoding a putative PIP_2-DAG lipase. The electroantennogram responses to odor stimulation are strongly reduced in these mutants.[72] Taken together, any disturbance of the PIP_2 cleavage and regeneration cycle seems to impair an appropriate processing of the olfactory signal in the receptor neurons.

That OR channels also conduct Ca^{2+} provides the basis for an amplification of odor signals by intracellular Ca^{2+} release, via activation of ryanodine receptors. Inhibition of ryanodine receptors abolished the observed signal amplification.[69] Signal amplification is only possible until a certain signal strength or duration is achieved. Long-lasting odor stimulation or high odor concentration leads to the contrary, i.e., OSN adaptation.[28] In Ca^{2+}-conducting proteins, the Ca^{2+} influx often regulates the channel dynamics via calmodulin (CaM). In voltage-gated Ca^{2+} channels and in CNG channels, CaM activation due to Ca^{2+} influx leads to channel closure to protect the cell from Ca^{2+} overload.[73] In line with this, the Ca^{2+} extrusion rate of the plasma membrane Ca^{2+} pump is largely enhanced by activated CaM.[74,75] For *Drosophila* Orco channels, CaM activity enhances the ion flux via Orco.[76] This positive feedback might play a role in the Orco activity-dependent OR sensitization.[60] For the function of ORs, the specific effect of CaM depends on the OrX proteins and may vary.[76] For Or22a and Or56a, CaM enhances the OR response, and for Or33a and Or47a, there is no effect.

A new kind of OR regulation was recently discovered.[77] A proper *Drosophila* OR function requires a functional phospholipid flippase, dATP8B, that maintains the asymmetry in phospholipid distribution across the plasma membrane. Flies with mutation in the dATP8B transporter show a largely diminished sensitivity in ORs, but interestingly not in IRs or GRs. As the transport of OR proteins to the dendritic membrane of OSNs is not affected by mutations in dATP8B, it is assumed that the transporter function is necessary for normal OR functionality once inserted into the membrane.[77] By contrast, for the pheromone receptor Or67d, a mislocalization was observed in dATP8B mutants.[78]

3. ROLE OF ORCO

Orco proteins are highly conserved within winged insects.[79] In wingless *Thermobia* (Thysanura), Orco-like proteins are expressed without OrX proteins.[27] It remains to be seen how these proteins might work. As OR proteins they are expected to detect odors.

In Orco mutant flies, there is no insertion of OrX proteins into the OSN membrane, i.e., Orco has a chaperone function.[17] The requirement of Orco activation for OR sensitization was already mentioned above.[60] A possible role of Orco as pacemaker channel was discussed[80] and observed for pheromone signaling in *Manduca*.[58] For *Drosophila*, this view could not be supported. Injection of the membrane-permeable cAMP analog, 8-bromo-cAMP, which activates Orco channels but not ORs, did not enhance the OSN background activity.[60] However, injection of the Or22a ligand ethyl butyrate increased the spontaneous OSN activity.

Orco can be activated by cyclic nucleotides[38] and by phosphorylation via PKC.[64] This allows the Orco channel to operate via the $G\alpha_s$ protein pathway or via the $G\alpha_q$ pathway. Another regulator is Ca^{2+}; very low cytoplasmic concentration inhibits enzymes such as PLC and PKC which causes dephosphorylation of Orco and thus inactivity. An increase in Ca^{2+} concentration activates CaM and stimulates Orco function.[76] There are synthetic Orco agonists which are related to VUAA1, the first compound discovered to act as allosteric agonist of OR complexes.[50–52] A couple of VUAA1 derivatives act as Orco inhibitors.[51] Further inhibitors are the amiloride derivatives

HMA and MIA.[81,82] Another potent inhibitor is the cation channel blocker ruthenium red.[37] A quite recent observation was the block of OR and Orco signaling by KB-R7943, an inhibitor of Na^+/Ca^{2+} exchange.[83]

4. FINAL REMARKS

It is interesting to compare chemosensory organs with a neuronal synapse.[84] Both systems receive and process chemical information. Thus, the molecule-detecting region, the dendrites of OSNs in insect olfaction, and the postsynaptic part of a synapse are equipped with chemoreceptors. These might be quite similar, as seen for iGluRs in the synapse and the iGluR-related IRs in OSNs. At the postsynaptic membrane, receptors are often linked to scaffolding proteins that connect signaling partners. In *Drosophila* photoreceptors, the mediators of the signal transduction cascade are connected by the scaffolding protein INAD.[85] It remains to be seen whether olfactory receptors form similar assemblies. The processing of olfactory information is linked to the physiological and intentional state of an individual.[43,57,86] Fine tuning of olfactory information processing is, for example, mediated by neuropeptides.[87] In summary, the detection and processing of volatile information in insects involve ionotropic as well as metabotropic mechanisms. The choice of a mechanism may depend on the concentration of the olfactory cue as seen for the switch in carbon dioxide detection from GRs to IRs with increasing concentration. As stated above, many other questions remain elusive such as topology and stoichiometry of OR channels. In addition, there is still a lack in understanding the mechanisms that tune an olfactory system to become as sensitive as a pheromone receptor. Further progress in the development of new techniques will likely provide answers to these open questions. New optogenetic approaches, for example, make it possible to control cAMP production via the light-sensitive adenylyl cyclase PAC α in individual OSNs.[88] Such noninvasive tools will also allow us to determine the role of individual olfactory receptors for odor-guided behavior. When OSNs expressing a given receptor are equipped with channelrhodopsin-2, they can be specifically activated by blue light which in turn provides information on their impact in neuronal networks.[88]

ACKNOWLEDGMENTS

This study was supported by the Max Planck Society. The author thanks Merid Getahun and Richard Glatz for helpful comments on the manuscript.

REFERENCES

1. Bargmann CI. Comparative chemosensation from receptors to ecology. *Nature.* 2006;444(7117):295–301.
2. Touhara K, Vosshall LB. Sensing odorants and pheromones with chemosensory receptors. *Annu Rev Physiol.* 2009;71:307–332.
3. Kaupp UB. Olfactory signalling in vertebrates and insects: differences and commonalities. *Nat Rev Neurosci.* 2010;11(3):188–200.
4. Benton R. Chemical sensing in *Drosophila. Curr Opin Neurobiol.* 2008;18(4):357–363.
5. Montell C. A taste of the *Drosophila* gustatory receptors. *Curr Opin Neurobiol.* 2009;19(4):345–353.
6. Leal WS. Odorant reception in insects: roles of receptors, binding proteins, and degrading enzymes. *Annu Rev Entomol.* 2013;58:373–391.
7. Galizia CG. Olfactory coding in the insect brain: data and conjectures. *Eur J Neurosci.* 2014;39(11):1784–1795.
8. Galizia CG, Rössler W. Parallel olfactory systems in insects: anatomy and function. *Annu Rev Entomol.* 2010;55:399–420.
9. Hallem EA, Carlson JR. Coding of odors by a receptor repertoire. *Cell.* 2006;125(1):143–160.
10. Semmelhack JL, Wang JW. Select *Drosophila* glomeruli mediate innate olfactory attraction and aversion. *Nature.* 2009;459(7244):218–223.
11. Stensmyr MC, Dweck HK, Farhan A, et al. A conserved dedicated olfactory circuit for detecting harmful microbes in *Drosophila. Cell.* 2012;151(6):1345–1357.
12. Hansson BS, Knaden M, Sachse S, Stensmyr MC, Wicher D. Towards plant-odor-related olfactory neuroethology in *Drosophila. Chemoecology.* 2010;20:51–61.
13. Hansson BS, Stensmyr MC. Evolution of insect olfaction. *Neuron.* 2011;72(5):698–711.
14. Berg HC, Purcell EM. Physics of chemoreception. *Biophys J.* 1977;20(2):193–219.
15. Wicher D. Functional and evolutionary aspects of chemoreceptors. *Front Cell Neurosci.* 2012;6:48.
16. Reisert J, Bauer PJ, Yau KW, Frings S. The Ca-activated Cl channel and its control in rat olfactory receptor neurons. *J Gen Physiol.* 2003;122(3):349–363.
17. Larsson MC, Domingos AI, Jones WD, Chiappe ME, Amrein H, Vosshall LB. Or83b encodes a broadly expressed odorant receptor essential for *Drosophila* olfaction. *Neuron.* 2004;43(5):703–714.
18. Neuhaus EM, Gisselmann G, Zhang W, Dooley R, Störtkuhl K, Hatt H. Odorant receptor heterodimerization in the olfactory system of *Drosophila melanogaster. Nat Neurosci.* 2005;8(1):15–17.
19. Robertson HM, Kent LB. Evolution of the gene lineage encoding the carbon dioxide receptor in insects. *J Insect Sci.* 2009;9(19):19.
20. Jones WD, Cayirlioglu P, Kadow IG, Vosshall LB. Two chemosensory receptors together mediate carbon dioxide detection in *Drosophila. Nature.* 2007;445(7123):86–90.
21. Kwon JY, Dahanukar A, Weiss LA, Carlson JR. The molecular basis of CO_2 reception in *Drosophila. Proc Natl Acad Sci USA.* 2007;104(9):3574–3578.
22. Benton R, Sachse S, Michnick SW, Vosshall LB. Atypical membrane topology and heteromeric function of *Drosophila* odorant receptors in vivo. *PLoS Biol.* 2006;4(2):e20.
23. Lundin C, Kall L, Kreher SA, et al. Membrane topology of the *Drosophila* OR83b odorant receptor. *FEBS Lett.* 2007;581(29):5601–5604.
24. Abuin L, Bargeton B, Ulbrich MH, Isacoff EY, Kellenberger S, Benton R. Functional architecture of olfactory ionotropic glutamate receptors. *Neuron.* 2011;69(1):44–60.
25. Croset V, Rytz R, Cummins SF, et al. Ancient protostome origin of chemosensory ionotropic glutamate receptors and the evolution of insect taste and olfaction. *PLoS Genet.* 2010;6(8):e1001064.

26. Penalva-Arana DC, Lynch M, Robertson HM. The chemoreceptor genes of the water-flea *Daphnia pulex*: many Grs but no Ors. *BMC Evol Biol.* 2009;9(79):79.
27. Missbach C, Dweck HK, Vogel H, et al. Evolution of insect olfactory receptors. *Elife.* 2014;3(3):e02115.
28. Getahun MN, Wicher D, Hansson BS, Olsson SB. Temporal response dynamics of *Drosophila* olfactory sensory neurons depends on receptor type and response polarity. *Front Cell Neurosci.* 2012;6:54.
29. Rytz R, Croset V, Benton R. Ionotropic receptors (IRs): chemosensory ionotropic glutamate receptors in *Drosophila* and beyond. *Insect Biochem Mol Biol.* 2013;43(9):888–897.
30. Badsha F, Kain P, Prabhakar S, et al. Mutants in *Drosophila* TRPC channels reduce olfactory sensitivity to carbon dioxide. *PLoS One.* 2012;7(11):e49848.
31. Yao CA, Carlson JR. Role of G-proteins in odor-sensing and CO_2-sensing neurons in *Drosophila. J Neurosci.* 2010;30(13):4562–4572.
32. Ai M, Min S, Grosjean Y, et al. Acid sensing by the *Drosophila* olfactory system. *Nature.* 2010;468(7324):691–695.
33. Boto T, Gomez-Diaz C, Alcorta E. Expression analysis of the 3 G-protein subunits, Galpha, Gbeta, and Ggamma, in the olfactory receptor organs of adult *Drosophila melanogaster. Chem Senses.* 2010;35(3):183–193.
34. Wetzel CH, Behrendt HJ, Gisselmann G, Störtkuhl KF, Hovemann B, Hatt H. Functional expression and characterization of a *Drosophila* odorant receptor in a heterologous cell system. *Proc Natl Acad Sci USA.* 2001;98(16):9377–9380.
35. Grosse-Wilde E, Svatos A, Krieger J. A pheromone-binding protein mediates the bombykol-induced activation of a pheromone receptor in vitro. *Chem Senses.* 2006;31(6):547–555.
36. Nordström KJ, Almen MS, Edstam MM, Fredriksson R, Schioth HB. Independent HHsearch, Needleman-Wunsch-based and motif analyses reveals the overall hierarchy for most of the G protein-coupled receptor families. *Mol Biol Evol.* 2011;28:2471–2480.
37. Sato K, Pellegrino M, Nakagawa T, Nakagawa T, Vosshall LB, Touhara K. Insect olfactory receptors are heteromeric ligand-gated ion channels. *Nature.* 2008;452:1002–1006.
38. Wicher D, Schäfer R, Bauernfeind R, et al. *Drosophila* odorant receptors are both ligand-gated and cyclic-nucleotide-activated cation channels. *Nature.* 2008;452:1007–1011.
39. Nakagawa T, Vosshall LB. Controversy and consensus: noncanonical signaling mechanisms in the insect olfactory system. *Curr Opin Neurobiol.* 2009;19(3):284–292.
40. Müller M, Bamann C, Bamberg E, Kühlbrandt W. Projection structure of channelrhodopsin-2 at 6 Å resolution by electron crystallography. *J Mol Biol.* 2011;414(1):86–95.
41. Silbering AF, Benton R. Ionotropic and metabotropic mechanisms in chemoreception: 'chance or design'? *EMBO Rep.* 2010;11(3):173–179.
42. Martin F, Boto T, Gomez-Diaz C, Alcorta E. Elements of olfactory reception in adult *Drosophila melanogaster. Anat Rec (Hoboken).* 2013;296(9):1477–1488.
43. Sakurai T, Namiki S, Kanzaki R. Molecular and neural mechanisms of sex pheromone reception and processing in the silkmoth *Bombyx mori. Front Physiol.* 2014;5(125):125.
44. Bohbot JD, Dickens JC. Selectivity of odorant receptors in insects. *Front Cell Neurosci.* 2012;6:29.
45. Dekker T, Ibba I, Siju KP, Stensmyr MC, Hansson BS. Olfactory shifts parallel superspecialism for toxic fruit in *Drosophila melanogaster* sibling, *D. sechellia. Curr Biol.* 2006;16(1):101–109.
46. Kato HE, Zhang F, Yizhar O, et al. Crystal structure of the channelrhodopsin light-gated cation channel. *Nature.* 2012;482(7385):369–374.
47. Penna A, Demuro A, Yeromin AV, et al. The CRAC channel consists of a tetramer formed by Stim-induced dimerization of Orai dimers. *Nature.* 2008;456(7218):116–120.

48. Pask GM, Jones PL, Rützler M, Rinker DC, Zwiebel LJ. Heteromeric anopheline odorant receptors exhibit distinct channel properties. *PLoS One*. 2011;6(12):e28774.
49. Nakagawa T, Pellegrino M, Sato K, Vosshall LB, Touhara K. Amino acid residues contributing to function of the heteromeric insect olfactory receptor complex. *PLoS One*. 2012;7(3):e32372.
50. Jones PL, Pask GM, Rinker DC, Zwiebel LJ. Functional agonism of insect odorant receptor ion channels. *Proc Natl Acad Sci USA*. 2011;108(21):8821–8825.
51. Chen S, Luetje CW. Identification of new agonists and antagonists of the insect odorant receptor co-receptor subunit. *PLoS One*. 2012;7(5):e36784.
52. Taylor RW, Romaine IM, Liu C, et al. Structure-activity relationship of a broad-spectrum insect odorant receptor agonist. *ACS Chem Biol*. 2012;7(10):1647–1652.
53. German PF, van der Poel S, Carraher C, Kralicek AV, Newcomb RD. Insights into subunit interactions within the insect olfactory receptor complex using FRET. *Insect Biochem Mol Biol*. 2013;43(2):138–145.
54. Smart R, Kiely A, Beale M, et al. *Drosophila* odorant receptors are novel seven transmembrane domain proteins that can signal independently of heterotrimeric G proteins. *Insect Biochem Mol Biol*. 2008;38(8):770–780.
55. Stengl M. Inositol-trisphosphate-dependent calcium currents precede cation currents in insect olfactory receptor neurons in vitro. *J Comp Physiol A*. 1994;174(2):187–194.
56. Krieger J, Breer H. Olfactory reception in invertebrates. *Science*. 1999;286(5440): 720–723.
57. Stengl M. Pheromone transduction in moths. *Front Cell Neurosci*. 2010;4:133.
58. Nolte A, Funk NW, Mukunda L, et al. In situ tip-recordings found no evidence for an Orco-based ionotropic mechanism of pheromone-transduction in *Manduca sexta*. *PLoS One*. 2013;8(5):e62648.
59. Mukunda L, Lavista Llanos S, Hansson BS, Wicher D. Dimerisation of the *Drosophila* odorant coreceptor Orco. *Front Cell Neurosci*. 2014;8:261.
60. Getahun MN, Olsson SB, Lavista-Llanos S, Hansson BS, Wicher D. Insect odorant response sensitivity is tuned by metabotropically autoregulated olfactory receptors. *PLoS One*. 2013;8(3):e58889.
61. Benton R, Vannice KS, Vosshall LB. An essential role for a CD36-related receptor in pheromone detection in *Drosophila*. *Nature*. 2007;450(7167):289–293.
62. Laughlin JD, Ha TS, Jones DN, Smith DP. Activation of pheromone-sensitive neurons is mediated by conformational activation of pheromone-binding protein. *Cell*. 2008;133(7):1255–1265.
63. Gomez-Diaz C, Reina JH, Cambillau C, Benton R. Ligands for pheromone-sensing neurons are not conformationally activated odorant binding proteins. *PLoS Biol*. 2013;11(4):e1001546.
64. Sargsyan V, Getahun MN, Lavista Llanos S, Olsson SB, Hansson BS, Wicher D. Phosphorylation via PKC regulates the function of the *Drosophila* odorant coreceptor. *Front Cell Neurosci*. 2011;5:5.
65. Martin F, Charro MJ, Alcorta E. Mutations affecting the cAMP transduction pathway modify olfaction in *Drosophila*. *J Comp Physiol A*. 2001;187(5):359–370.
66. Gomez-Diaz C, Martin F, Alcorta E. The cAMP transduction cascade mediates olfactory reception in *Drosophila melanogaster*. *Behav Genet*. 2004;34(4):395–406.
67. Deng Y, Zhang W, Farhat K, Oberland S, Gisselmann G, Neuhaus EM. The stimulatory Galpha(s) protein is involved in olfactory signal transduction in *Drosophila*. *PLoS One*. 2011;6(4):e18605.
68. Chatterjee A, Roman G, Hardin PE. Go contributes to olfactory reception in *Drosophila melanogaster*. *BMC Physiol*. 2009;9(22):22.
69. Ignatious Raja JS, Katanayeva N, Katanaev VL, Galizia CG. Role of Go/i subgroup of G proteins in olfactory signaling of *Drosophila melanogaster*. *Eur J Neurosci*. 2014;39(8):1245–1255.

70. Riesgo-Escovar J, Raha D, Carlson JR. Requirement for a phospholipase C in odor response: overlap between olfaction and vision in *Drosophila*. *Proc Natl Acad Sci USA*. 1995;92(7):2864–2868.
71. Kain P, Chakraborty TS, Sundaram S, Siddiqi O, Rodrigues V, Hasan G. Reduced odor responses from antennal neurons of G(q)alpha, phospholipase Cbeta, and rdgA mutants in *Drosophila* support a role for a phospholipid intermediate in insect olfactory transduction. *J Neurosci*. 2008;28(18):4745–4755.
72. Kain P, Chandrashekaran S, Rodrigues V, Hasan G. *Drosophila* mutants in phospholipid signaling have reduced olfactory responses as adults and larvae. *J Neurogenet*. 2009;23(3):303–312.
73. Trudeau MC, Zagotta WN. Calcium/calmodulin modulation of olfactory and rod cyclic nucleotide-gated ion channels. *J Biol Chem*. 2003;278(21):18705–18708.
74. Budde T, Meuth S, Pape HC. Calcium-dependent inactivation of neuronal calcium channels. *Nat Rev Neurosci*. 2002;3(11):873–883.
75. Strehler EE, Filoteo AG, Penniston JT, Caride AJ. Plasma-membrane Ca(2+) pumps: structural diversity as the basis for functional versatility. *Biochem Soc Trans*. 2007;35(Pt 5):919–922.
76. Mukunda L, Miazzi F, Kaltofen S, Hansson BS, Wicher D. Calmodulin modulates insect odorant receptor function. *Cell Calcium*. 2014;55:324–333.
77. Liu YC, Pearce MW, Honda T, et al. The *Drosophila melanogaster* phospholipid flippase dATP8B is required for odorant receptor function. *PLoS Genet*. 2014;10(3):e1004209.
78. Ha TS, Xia R, Zhang H, Jin X, Smith DP. Lipid flippase modulates olfactory receptor expression and odorant sensitivity in *Drosophila*. *Proc Natl Acad Sci USA*. 2014;111(21):7831–7836.
79. Vosshall LB, Hansson BS. A unified nomenclature system for the insect olfactory co-receptor. *Chem Senses*. 2011;36:497–498.
80. Stengl M, Funk NW. The role of the coreceptor Orco in insect olfactory transduction. *J Comp Physiol A*. 2013;199(11):897–909.
81. Pask GM, Bobkov YV, Corey EA, Ache BW, Zwiebel LJ. Blockade of insect odorant receptor currents by amiloride derivatives. *Chem Senses*. 2013;38(3):221–229.
82. Rollecke K, Werner M, Ziemba PM, Neuhaus EM, Hatt H, Gisselmann G. Amiloride derivatives are effective blockers of insect odorant receptors. *Chem Senses*. 2013;38(3):231–236.
83. Bobkov Y, Corey E, Ache B. An inhibitor of Na(+)/Ca(2+) exchange blocks activation of insect olfactory receptors. *Biochem Biophys Res Commun*. 2014;450(2):1104–1109.
84. Shaham S. Chemosensory organs as models of neuronal synapses. *Nat Rev Neurosci*. 2010;11(3):212–217.
85. Katz B, Minke B. *Drosophila* photoreceptors and signaling mechanisms. *Front Cell Neurosci*. 2009;3:2.
86. Dweck HK, Ebrahim SA, Kromann S, et al. Olfactory preference for egg laying on citrus substrates in *Drosophila*. *Curr Biol*. 2013;23(24):2472–2480.
87. Ignell R, Root CM, Birse RT, Wang JW, Nässel DR, Winther AM. Presynaptic peptidergic modulation of olfactory receptor neurons in *Drosophila*. *Proc Natl Acad Sci USA*. 2009;106(31):13070–13075.
88. Störtkuhl KF, Fiala A. The smell of blue light: a new approach toward understanding an olfactory neuronal network. *Front Neurosci*. 2011;5(72):72.

CHAPTER THREE

Advances in the Identification and Characterization of Olfactory Receptors in Insects

Nicolas Montagné*, Arthur de Fouchier†, Richard D. Newcomb‡,§, Emmanuelle Jacquin-Joly†,1

*Institute of Ecology & Environmental Sciences of Paris, UPMC-Sorbonne Universités, Paris, France
†Institute of Ecology & Environmental Sciences of Paris, INRA, Versailles, France
‡School of Biological Sciences, University of Auckland, Auckland, New Zealand
§The New Zealand Institute for Plant & Food Research Limited, Auckland, New Zealand
[1]Corresponding author: e-mail address: emmanuelle.jacquin@versailles.inra.fr

Contents

Abstract

Olfactory receptors (ORs) are the key elements of the molecular machinery responsible for the detection of odors in insects. Since their initial discovery in *Drosophila melanogaster* at the beginning of the twenty-first century, insect ORs have been the focus of intense research, both for fundamental knowledge of sensory systems and for their potential as novel targets for the development of products that could impact harmful behaviors of crop pests and disease vectors. In recent years, studies on insect ORs have entered the genomic era, with an ever-increasing number of OR genes being characterized every year through the sequencing of genomes and transcriptomes. With the upcoming release of genomic sequences from hundreds of insect species, the insect OR family could very well become the largest multigene family known. This extremely rapid identification of ORs in many insects is driving the necessity for the development

Progress in Molecular Biology and Translational Science, Volume 130
ISSN 1877-1173
http://dx.doi.org/10.1016/bs.pmbts.2014.11.003

of high-throughput technologies that will allow the identification of ligands for this unprecedented number of receptors. Moreover, such technologies will also be important for the development of agonists or antagonists that could be used in the fight against pest insects.

ABBREVIATIONS

EST expressed sequence tag
GPCR G-protein-coupled receptor
HEK human embryonic kidney
HT high-throughput
NGS next-generation sequencing
OBP odorant-binding protein
ODE odorant-degrading enzyme
OR olfactory receptor
Orco olfactory receptor coreceptor
OSN olfactory sensory neuron
SNMP sensory neuron membrane protein
WGS whole-genome shotgun

1. INTRODUCTION: THE MOLECULAR BASES OF ODOR DETECTION IN INSECTS

Today's research in biology is becoming more interdisciplinary with the integration of complementary fields, from genomics to ecology. Research on sensory systems such as the insect olfactory system is no exception, providing the opportunity to investigate the interactions between the genome, neurobiological processes, behaviors, and the environment of an insect. Olfaction is key for insects, since vital behaviors such as foraging, predation and pathogen avoidance, mate recognition, and oviposition site selection are primarily dependent on the detection of blends of airborne molecules either emitted by conspecifics (pheromones) or by other organisms (kairomones).[1]

Olfactory appendages of insects consist of antennae and—to a lesser extent—maxillary palps. Although these appendages have a large variety of shapes and sizes, depending on the evolutionary history and the ecology of insects, they are always covered with minute cuticular extensions called sensilla, which can take different morphological forms[2] (Fig. 1). Olfactory sensory neurons (OSNs) responsible for the detection of volatile molecules are housed within these sensilla, where they are bathed in an aqueous

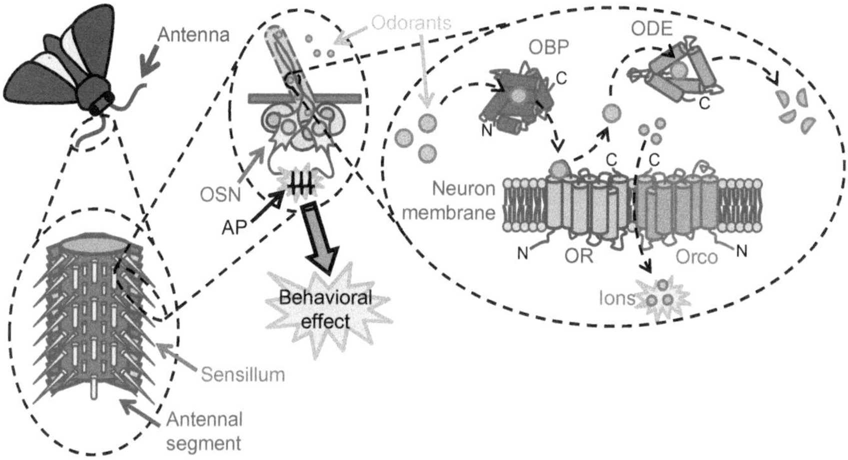

Figure 1 Illustration of a lepidopteran antenna and the different steps involved in odorant reception within an olfactory sensillum. After entering the sensillum lymph through cuticular pores, odorant molecules are presumably carried by odorant-binding proteins (OBPs) to cross the lymph. Then, they reach the dendritic membrane of the olfactory sensory neuron (OSN), where they interact with a heteromeric receptor formed by an olfactory receptor (OR) and its coreceptor (Orco). OR–Orco activation leads to the generation of action potentials (APs) transmitted to the antennal lobes in the central nervous system. Signal termination is thought to be conducted by enzymatic degradation of the odorant molecules via odorant-degrading enzymes (ODEs).

environment, the sensillar lymph. Insects possess various numbers of OSN functional classes (from dozens to hundreds), detecting different—yet often overlapping—subsets of odorants.[3–5] Several large multigene families are involved in the molecular processes that take place at the level of these OSNs, allowing the transduction of chemical information into electrical signals that can be interpreted by the insect's brain (Fig. 1). The main steps of this process at the molecular level are the transport of volatile molecules within the sensillar lymph, the binding of these molecules to receptors expressed at the membrane of OSNs, and the degradation of the signal molecules following receptor activation. Whereas some hydrophilic odorants (acids or amines) can dissolve readily within an aqueous environment, most volatile molecules are lipophilic (aromatics, terpenes, straight-chain alcohols, esters, etc.) and they are believed to be bound and transported within this aqueous environment by odorant-binding proteins (OBPs) and chemosensory proteins that are present at high concentrations within the lymph (for a review, see Ref. 6). The sensillar lymph also contains many hydrolytic enzymes, which are suspected to play an important role in the degradation of odorant molecules, allowing the termination of the response to avoid overstimulation of the OSNs.[7–9]

In insects, membrane receptors that detect odorants belong to two different families. The ionotropic receptors are ancient chemosensory receptors that emerged during the evolution of protostomes, which includes arthropods, roundworms, molluscs, and annelid worms.[10,11] Insects possess a low number of such receptors (typically between 10 and 30) that are likely to be specialized in the detection of waterborne molecules. In *Drosophila*, they are known to bind acids, amines and aldehydes.[12]

The second family of chemosensory membrane receptors is the olfactory receptor (OR) family. They are specific to insects, probably having evolved from gustatory receptors in a common ancestor of winged insects.[13,14] They can bind various lipophilic airborne molecules, for example, aromatics (very common in flower odors), terpenes (emitted in high quantities by flowering plants and coniferous), and fatty acid derivatives (constituents of many sex pheromones).[15–18] Insect ORs form functional complexes with a unique coreceptor (named Orco) that is also required for correct targeting of the OR proteins to the dendritic membrane.[19,20] Within the dendrite, the OR–Orco complex forms a ligand-gated nonselective cation channel that can open directly upon activation by an appropriate ligand.[21–23]

ORs are responsible for the ability to detect the majority of odors by a given insect, albeit their number is highly variable between insect groups. For example, the genome of the body louse, *Pediculus humanus*, harbors only 10 OR genes,[24] whereas genomes of the ants, *Camponotus floridanus* and *Harpegnathos saltator*, possess 350 candidate OR genes.[25] Each OSN generally expresses only a single OR gene together with Orco,[26] with the functional properties of the OR responsible for the receptive range of the OSN.

2. IDENTIFICATION OF COMPLETE INSECT OR REPERTOIRES COULD ONLY BE ACHIEVED BY GENOME AND TRANSCRIPTOME SEQUENCING

The first receptor proteins responsible for the detection of odors in an animal were identified in the rat in 1991,[27] a discovery recognized by the Nobel Prize in Physiology or Medicine in 2004.[28] Subsequently, a wide range of other vertebrate ORs have been identified in various species including humans,[29] fish,[30] and birds,[31] revealing strong conservation across the chordates.[32] Homology-based strategies were conducted in insects by many researchers at that time to identify candidate insect ORs, but remained unsuccessful. For example, attempts to homology-clone bee ORs with degenerate oligonucleotides derived from mammalian sequences only amplified vertebrate contaminants.[33] The field had to wait for the first insect genome sequencing project—that of the vinegar fly *Drosophila melanogaster* in 2000[34]—to discover the first insect ORs (see Section 2.2). This revealed that insect ORs while containing seven-transmembrane regions were not G-protein-coupled receptors (GPCRs) like mammalian ORs and were highly divergent at the sequence level, thus making it difficult to isolate them—if not impossible—by homology cloning strategies. The rapid progresses in sequencing technologies and associated bioinformatic tools now allowed the rapid identification of ORs in any species of interest and at a moderate cost.

2.1. Advances in sequencing technologies and bioinformatic tools

In the early 2000s, sequencing a large and complex eukaryotic genome was challenging and expensive. The annotated sequence of the *D. melanogaster* 180 Mb genome resulted from both public and private funds and from the concerted efforts of ∼200 experimental and computational biologists from more than 30 institutions in eight countries.[34] At that time, it was the first application of the whole-genome shotgun (WGS) approach, based on Sanger sequencing,[35] to sequence an entire animal genome. The WGS approach has been extended to the sequencing of other insect genomes, such as the mosquito *Anopheles gambiae*,[36] the silkworm *Bombyx mori*,[37,38] the European honeybee *Apis mellifera*,[39] the red flour beetle *Tribolium castaneum*,[40] and more recently the human body louse.[24]

Sanger sequencing, or chain termination method, has more recently been supplanted by so called next-generation sequencing or "NGS" methods that produce far greater volumes of sequence data by sequencing large numbers of DNA molecules simultaneously. These technologies, which include 454/Roche FLX, Illumina/Solexa and ABI/SOLiD, offer many advantages such as low-cost per base sequenced, the ability to use low quantity templates, short run times, and vast data volumes. For a detailed review of the different sequencing technologies, see Ref. 41. Many recent insect genome sequencing efforts, including butterflies *Danaus plexippus*[42] and *Heliconius melpomene*,[43] and the diamondback moth *Plutella xylostella*,[44] relied on these newer technologies.

The large quantities of data produced by NGS have required the parallel development of new bioinformatics algorithms and programs for the analysis of sequence data. However, sophisticated analyses often rely on programming and database skills and the application of bioinformatic tools is often a stumbling block for most experimental biologists with no programming experience. The development of new platforms such as Galaxy[45] opens the way for high-throughput (HT) data analyses and NGS manipulation.[46] This open, web-based platform for data-intensive biology is usable by most researchers, even without a background in bioinformatics, and the rapid development of Galaxy servers will undoubtedly lead to an increase in the ability of biologists to analyze and interpret data.

2.2. Exploitation of insect genomes for OR identification

As discussed above, the first animal ORs were not identified in an insect, but in the rat.[27] Since it was known at that time that olfactory signal transduction in rat olfactory neurons involved cAMP (thus probably heterotrimeric G-proteins), Linda Buck and Richard Axel looked for—and found—a subfamily of multiple and divergent genes encoding GPCRs that were expressed specifically in the olfactory epithelium. Following this initial discovery and benefiting from the on-going sequencing of the *D. melanogaster* genome, three research groups applied a similar strategy combining bioinformatics and molecular biology for the discovery of ORs in insects. The genomic data available at that time were mined for gene regions encoding predicted multi-transmembrane domain proteins, and those whose expression was spatially restricted to some OSNs in the antennae and maxillary palps were retained as OR candidates.[47–49] This initial dozen or so candidate genes allowed the subsequent identification of the entire OR repertoire of the vinegar fly by sequence similarity. The OR multigene family of *D. melanogaster* is composed of 61 genes encoding 62 OR proteins.[13] Some years later, it was finally demonstrated that insect ORs, first assumed to be GPCRs, actually belong to a distinct class of seven-transmembrane domain receptors, that have an inverted topology compared to GPCRs and can signal independently of G-proteins.[20–23]

Interestingly, even with the *Drosophila* OR gene sequences available, it was not immediately possible to homology-amplify other insect ORs by PCR, except the conserved coreceptor Orco.[50–52] Many unsuccessful attempts were conducted in numerous laboratories, including ours, once the *Drosophila* OR sequences were available. The sequencing and analysis of the *A. gambiae* genome,[36] two years after that of *D. melanogaster*, revealed that *Anopheles* ORs constitute a family of 79 ORs very divergent from

Drosophila ORs, even though both species belong to the order Diptera.[53,54] How other insect ORs could be identified was through using the *Drosophila* ORs as queries in tBLASTn searches, as well as using the Quasi-Periodic Feature Classifier algorithm.[55]

Similar approaches were conducted on each newly sequenced insect genome, describing partial or complete repertoires of OR genes from a diversity of species from many orders including the Diptera, Hymenoptera, Lepidoptera, Coleoptera, Hemiptera, and Phtiraptera (summarized in Fig. 2A). As of July 2014, a total of 3145 insect candidate OR sequences have been identified through genome analyses (Table 1), representing

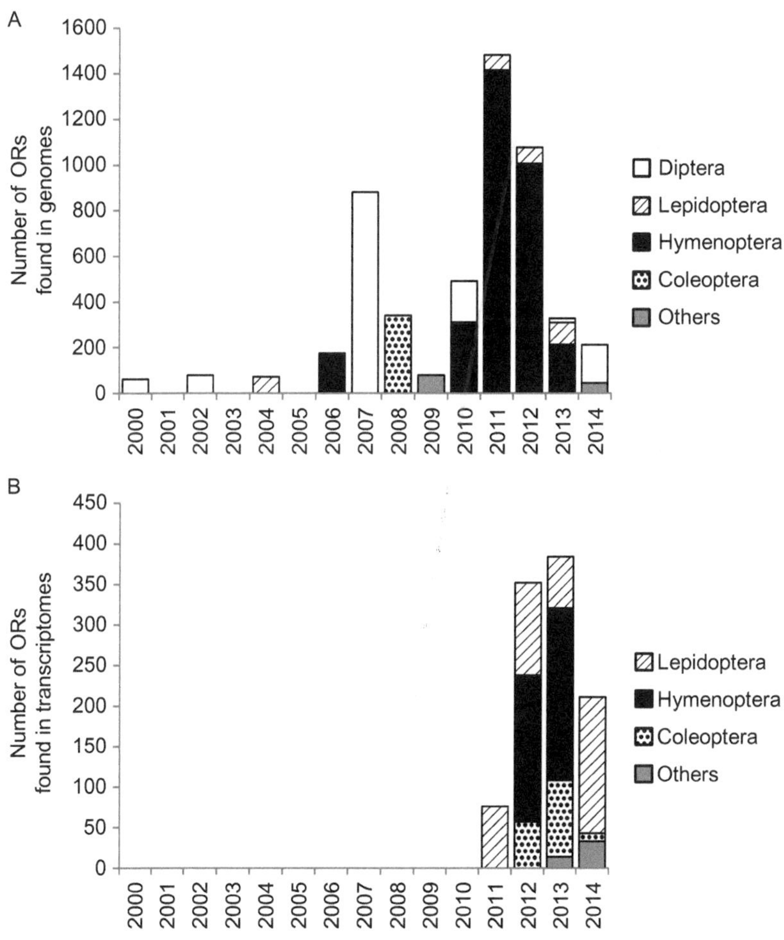

Figure 2 Evolution of the number of insect olfactory receptors identified through analysis of genomes (A) and transcriptomes (B).

Table 1 Olfactory receptor-encoding genes identified in insect genomes

Insect species	Number of OR genes	References
Coleoptera		
Tribolium castaneum	259	56
Diptera	**1287**	
Aedes aegypti	131	57
Anopheles darlingi	18	58
Anopheles gambiae	79	54
Culex quinquefasciatus	180	59
Drosophila ananassae	66	60
Drosophila erecta	61	60
Drosophila grimshawi	61	60
Drosophila melanogaster	61	13
Drosophila mojavensis	59	60
Drosophila persimilis	52	60
Drosophila pseudoobscura	64	60
Drosophila sechellia	54	60
Drosophila simulans	52	60
Drosophila virilis	53	60
Drosophila willistoni	65	60
Drosophila yakuba	63	60
Glossina morsitans	46	61
Mayetiola destructor	122	62
Hemiptera	**115**	
Acyrtosiphon pisum	79	146
Aphis gossypii	36	63
Hymenoptera	**2833**	
Acromyrmex echiniator	81	64
Apis florea	99	Unpublished
Apis mellifera	163	65

Table 1 Olfactory receptor-encoding genes identified in insect genomes—cont'd

Insect species	Number of OR genes	References
Atta cephalotes	215	66
Bombus impatiens	134	Unpublished
Bombus terrestris	116	Unpublished
Camponotus floridanus	352	25
Harpegnathos saltator	347	25
Linepithema humile	337	67
Megachile rotundata	123	Unpublished
Nasonia vitripennis	225	68
Pogonomyrmex barbatus	344	69
Solenopsis invicta	297	70
Lepidoptera	**302**	
Bombyx mori	71	17,43
Danaus plexippus	66	42
Heliconius melpomene	70	43
Plutella xylostella	95	71
Phthiraptera		
Pediculus humanus	10	24

one of the most abundant protein families within the public domain databases. As mentioned, these repertoires vary from 10 in the body louse[24] to 352 ORs in the carpenter ant[25] (Table 1), numbers that appear to be linked to the ecology of these species.

In the frame of the i5K international initiative,[72] whose objective is to sequence 5000 arthropod genomes in the coming years, we will soon have access to a large diversity of insect genomes to search for additional OR repertoires. More than 700 hexapod species have already been nominated for genome sequencing, with notably more than 250 hymenopterans, 100 dipterans, and 70 coleopterans.[73] Such endeavors will help scientists to understand the evolutionary origin of this atypical gene family and its expansion in various insect lineages. However, one has to keep in mind that OR repertoires predicted from genome analyses will overestimate the number of

ORs used by an insect at a specific developmental stage and in a certain tissue type. For instance, whereas the *B. mori* genome harbors 71 OR-encoding genes,[17,43,74] only 35 are expressed in adults, 24 in larvae and 25 could not be amplified by RT-PCR in either developmental stage.[17] Thus, these HT methods to identify ORs, based solely on bioinformatics, have limitations if one is interested in the molecular bases of the chemical ecology of an insect.

2.3. Developing insect antennal transcriptomes for OR identification

As an alternative to whole-genome sequencing, which still remains beyond the reach of many research groups, transcriptomics has also been used to identify chemosensory genes in insects. The advantage of transcriptomic approaches is that expressed genes can be sampled from a given tissue considering a given stage, rather than all genes within a genome. The first transcriptome sequencing attempts on insect antennae could only identify highly expressed transcripts, mainly OBPs, in the hawkmoth *Manduca sexta*.[75] With a similar approach, Whitfield *et al.*[76] discovered one potential OR from *A. mellifera*, that appeared to be the conserved Orco homolog. Later, Jordan *et al.*[77] identified candidate OR transcripts from expressed sequenced tags (ESTs) prepared from the tortricid moth, *Epiphyas postvittana*, and Patch *et al.*[78] discovered two others in *M. sexta*. However, since ORs are expressed at very low levels in the antennae, such a strategy often requires sequencing of several thousand ESTs to obtain only a few candidate OR sequences.

An alternative approach is to use normalized or subtractive libraries, in which rare transcripts are enriched. We used this strategy to demonstrate the power of such an approach for identifying divergent and low expressed genes in a complex tissue. Using Sanger sequencing, in partnership with the Genoscope in France, we sequenced more than 40,000 ESTs prepared from male and female antennae of the cotton leafworm, *Spodoptera littoralis*, and identified 36 ORs, including four candidate sex-pheromone receptors.[79,80]

Soon after, the development of the 454 pyrosequencing technology (see Section 2.1) heralded an increase in the number of sequenced antennal transcriptomes and in turn the number of ORs identified within each of these transcriptomes (Fig. 2B). Within the Lepidoptera, such analyses demonstrated some conservation in the number of OR genes expressed in adult antennae, with 40–50 ORs found in many (Table 2). It is likely that the entirety of the adult OR repertoires has not been identified using this method and that some additional genes are yet to be identified. Notwithstanding this, transcriptomics remains a relevant strategy for isolating ORs

Table 2 Insect olfactory receptors identified in transcriptomes

Insect species	Tissues	Number of ORs	Sequencing method	References
Coleoptera				
Agrilus planipennis	Antennae	2	454 pyrosequencing (140,000 reads)	81
Dastarcus helophoroides	Antennae	1	Illumina paired-end (51 M reads)	82
Dendroctonus ponderosae	Antennae	49	Sanger paired-end (>7600 ESTs) + 454 (>1,000,000 reads)	83
Ips typographus	Antennae	43	454 (350,000 reads) + Illumina (3.6 M reads)	83
Megacyllene caryae	Antennae	57	454 (>1,500,000 reads)	84
Monochamus alternatus	Antennae	9	Illumina paired-end (62 M reads)	82
Hymenoptera				
Apis cerrana cerana	Whole-body	18	Illumina paired-end (55 M reads)	85
Atta vollenweideri	Antennae	185	454 (normalized library) + Illumina	66
Cephus cinctus	Antennae	28	454 (500,000 reads)	86
Cotesia vestalis	Antennae	163	454 (>440,000 reads, 3 libraries)	87
Lepidoptera				
Agrotis ipsilon	Male and female antennae	42	454 (>1,000,000 reads, 2 libraries)	88
Chilo suppressalis	Male and female antennae	47	Illumina paired-end (115 M reads, 2 libraries)	89
Cydia pomonella	Male and female antennae	43	454 (>900,000 reads, 2 libraries)	90
Dendrolimus houi	Antennae	33	Illumina paired-end (32 M reads, normalized library)	91

Continued

Table 2 Insect olfactory receptors identified in transcriptomes—cont'd

Insect species	Tissues	Number of ORs	Sequencing method	References
Dendrolimus kikuchii	Antennae	33	Illumina paired-end (32 M reads, normalized library)	91
Helicoverpa armigera	Male and female antennae	47	454 (>1,200,000 reads, 2 libraries)	92
	Adult heads, tarsi and abdomens; larval antennae, fat body, and mouthparts	+13	Illumina paired-end (>240 M reads, 10 libraries incl. 2 normalized libraries)	93
Manduca sexta	Antennae	47	454 (275,000 reads, normalized library)	94
	Antennae, larval antennae and palps	+17	454 (normalized library)	95
Sesamia inferens	Antennae, pheromonal gland	39	Illumina paired-end (54 M reads)	96
Sesamia nonagrioides	Antennae, brain, ovipositor, larval antennae and palps	13	454 (>1,000,000 reads) + Illumina (315 M reads, 3 libraries)	97
Spodoptera littoralis	Male antennae	29	Sanger (>20,000 ESTs, normalized library)	79
	Female antennae	+7	Sanger (>18,000 ESTs, normalized library)	80
	Antennae, brain, proboscis, and larvae	+11	454 (>1,600,000 reads, 3 libraries) + Illumina (4.3 M reads, 2 libraries)	98
Neuroptera				
Chrysopa pallens	Whole-body	14	Illumina (58 M reads)	99
Phasmatodea				
Phyllium siccifolium	Antennae	<30	Illumina paired-end (30 M reads)	14
Zygentoma				
Thermobia domestica	Antennae	3	Illumina paired-end (27 M reads)	14

sequenced from a species in which sequencing of the genome has not been achieved, or is not possible (e.g., where genome size is prohibitively large).

Since 2013, all but a few published insect antennal transcriptomes have been sequenced using the Illumina technology (Table 2), an approach known as RNA-seq, which allows the quantification of gene expression in addition to the simple discovery of OR sequences. Such quantitative analyses are now commonly used, also in species with genomic data. For instance, in the tse-tse fly, *Glossina morsitans*, the analysis of RNA-seq data following the annotation of 46 ORs in the genome demonstrated that the expression of a single OR gene accounted for 90% of the total OR expression, thus suggesting a prominent role for this receptor in this disease vector.[61] Transcriptomic and genomic strategies are therefore becoming increasingly complementary for identifying ORs, and the combination of both may become a standard method for such studies in the future.

3. TOWARD THE DEVELOPMENT OF HIGH THROUGHPUT METHODS FOR THE FUNCTIONAL CHARACTERIZATION OF INSECT ORs

3.1. Description of the different methodologies

Although many candidate insect OR sequences are now available, most encoded proteins remain orphan; that is, their ligands remain unknown. In response to the need to functionally characterization these ORs, the scientific community has developed or adapted multiple expression techniques, both *in vitro* and *in vivo*, with varying HT capabilities. To date, these assay systems have allowed the deorphanization of some ORs in a variety of species, but only two large repertoires have been deorphanized, that of the two dipterans *D. melanogaster* and *A. gambiae*.[16,18,100–102]

3.1.1 In vitro *heterologous expression systems*

In order to deorphanize a receptor, *in vitro* expression strategies are often chosen, typically for their speed and convenience. Insect ORs are no exception. To date, the most abundantly used *in vitro* systems to study insect ORs are through heterologous expression in *Xenopus* oocytes and human embryonic kidney (HEK) cells. Some other *in vitro* systems have also been used for assaying OR function, including *Sf*9, S2, and High-Five insect cells, as well as most recently, cell-free expression systems.

The large size of *Xenopus* oocytes and the robustness of these cells, together with their metabolism, make them a suitable cell type for expressing large quantities of membrane receptors. As such, this approach was used to deorphanize the first insect OR, a *Drosophila* OR.[103] Today, it constitutes the most common technique used for heterologous expression and deorphanizing insect ORs, as illustrated in Fig. 3. The oocytes become ligand-responsive upon expression of an OR and Orco through injection of their RNA into the cell, and can be assessed for responsiveness one at a time, in a liquid medium, using the two-electrode voltage-clamp technique.[15,104] In this approach, the usually hydrophobic odors are dissolved in DMSO to achieve a stimulation of the OR-expressing oocytes.

Cultured cell lines, such as mammalian HEK293[105] and insect *Sf*9 cells,[106] are largely used for heterologous expression as they are straightforward to maintain in a laboratory and do not require animal rearing. HEK293 have been extensively used for OR expression and characterization in diverse species. OR-dependent excitation of these cells lines is tested with DMSO, methanol, or OBP-solubilized odorants using calcium imaging or patch-clamp techniques (among references, see Refs.[22,23,107–113]). Some of the attractions of using these cell lines are the ability to make stably transformed lines that express Orco and an OR on addition of an inducing agent, and compatibility with plate reader formats.[112]

The published use of *Sf*9 cells, on the other hand, remains less common,[21,114–117] even though these cells basally express an endogenous

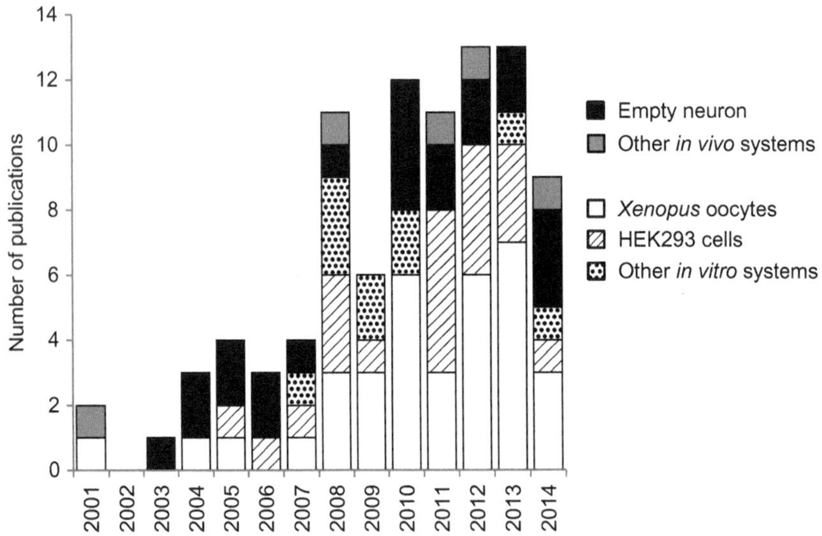

Figure 3 Number of publications per year reporting the use of the different methodologies described in Sections 3.1.1 and 3.1.2 for insect OR expression.

version of Orco.[21] The reduced uptake of the use of these cells is likely due to the low proportion of cells that respond to agonist, resulting in the need to use microscopy-based cell imaging techniques to monitor calcium flux. Other insect cells such as S2,[21,118] Bm5,[119] and High-Five cells[119,120] have been used only rarely for the expression of insect ORs. As well, mammalian HeLa expression is reported in one publication.[22]

Most recently, cell-free expression systems have been used to produce recombinant insect ORs, mainly for potential use in biosensor platforms. An insect cell-free expression system followed by integration into giant vesicles was used by a Japanese group.[121] Their proof-of-concept study utilized a sex-pheromone receptor of *B. mori* to demonstrate that a transmembrane receptor could be functionally produced in a cell-free manner. Similarly, expression of insect ORs has been achieved using the wheat-germ cell-free expression system, as well as in insect cells using the baculovirus expression system to produce relatively large quantities of membrane-bound insect ORs which can then be purified, for subsequent structural studies.[122] These new expression systems for insect ORs provide the potential for new functional approaches as well as providing an opportunity to study the unique structure of these receptor complexes.

3.1.2 In vivo *heterologous expression systems*

In vitro expression systems for OR deorphanization are convenient but suffer from the inability to factor in the contribution of the *in vivo* physiological environment of the receptors or produce the correct upstream processing. As such, *in situ* studies are still highly regarded in spite of being usually more difficult and time consuming. Thus, heterologous *in vivo* expression systems appear as a middle ground between *in vitro* and *in situ* expressions and a few of them have been developed for insect OR functional studies.

The most well known of these systems for insect ORs is the *Drosophila* "empty neuron" system. This is indeed an elegant expression strategy where a candidate OR is expressed in place of the endogenous DmelOR22a receptor in *D. melanogaster* antennae.[123] This is achieved through the association of a deletion and a particular combination of a GAL4 driver line employing the Or22a promotor and an UAS-controlled transgene. The response spectrum of the exogenous candidate OR can be investigated using the well-established single sensillum recording technique to monitor the electrical responses of the modified OSNs.

The interest in this approach is that OSNs offer the actual cellular machinery used to process ORs, allowing their proper trafficking and coupling with Orco. In addition, the odorants are delivered in the vapor phase, as in natural conditions. In this strategy, the candidate ORs are expressed in OSNs located in the basiconic ab3 sensilla of the flies, which offer an alcohol- and acetate-sensitive OSN sensillar environment.

Although this system appears useful for most ORs (see below), it may not be convenient for some specific families of insect ORs, such as lepidopteran pheromone receptors[124,125] and *Drosophila* ORs endogenously found in trichoid sensilla, such as the farnesol receptor DmelOR83c.[126] Thus, a variation of the "empty neuron" system has been developed to express ORs in *Drosophila* pheromone-tuned OSNs,[127] in which sensory neuron membrane protein 1 (SNMP1) is present. This SNMP may be necessary for the correct functioning of some insect ORs, including pheromone receptors.[126,128,129] In this system, through a similar GAL4/UAS combination, the candidate OR replaces DmelOR67d, the receptor for *cis*-vaccenyl acetate, in the *Drosophila* trichoid sensilla T1.

Apart from these two "empty neuron" expression systems, other *in vivo* expression strategies targeting other species/other organs have been developed. In 2011, a publication reported the use of the GAL4/UAS system to direct the expression of one of the *P. xylostella* candidate sex-pheromone receptors within *B. mori* pheromone-sensitive OSNs.[130] A behavioral assay allowed the identification of the receptor ligand as one of the main components of *P. xylostella* sex pheromone.

In *Drosophila*, a global overexpression strategy in antennae was published in 2001,[131] using a driver line that expresses Gal4 in OSNs of the third antennal segment and recording antennal responses via electroantennography. More recently, such an approach has been proved efficient for a non-dipteran OR,[125] using the promoter of Orco and the GAL4/UAS system to express a candidate OR in the majority of the OSNs of the fly antennae. In this approach, the global activity of the antennae is recorded via electroantennography and the response spectra of the candidate OR is highlighted by the modification in the antennal odorant response profile.[132] Again in *Drosophila*, ectopic expression of ORs in gustatory neurons[133] has also proved successful.

3.2. Large OR repertoire deorphanization

Among the different OR expression techniques described above, only *Xenopus* oocytes and "empty neuron" systems have been used to undertake large repertoire studies. The 2004 study by Hallem and Carlson[134] was the first of its kind. They used the ab3A "empty neuron" system, which had been established the year before,[123] to test 31 ORs of the fly with 10 odorants. The subsequent studies by the Carlson group increased the number of receptor and odorants studied[16,100,135] to 35 receptors, tested with up to 110 odorants. In *A. gambiae*, two studies were published at the same time by collaborating groups. One was conducted using the *Drosophila* "empty neuron",[18] and the other using *Xenopus* oocyte expression.[102] They succeeded in the deorphanization of 50 and 37 ORs, out of 72, using 110 and 88 compounds, respectively. The use of both *in vivo* and *in vitro* approaches for the same ORs allowed comparison of the two systems, revealing that most of the ORs investigated are functional in both systems and usually reveal similar response profiles *in vivo* and *in vitro*.[18,102] The overall conclusion of these studies is that ORs indeed determine the functional properties of the OSNs regarding basal firing rate, as well as their response spectrum, modalities, and dynamics. The theory of combinatorial coding of the olfactory information is supported by these publications, as odorants mostly activate multiple ORs. Yet, further studies will be needed to confirm that those different strategies give congruent results also for non-dipteran insects and that the selectivity and relative levels of sensitivity among ligands of heterologously expressed ORs indeed matches insect's OSNs with regards to these properties.

These studies were the first of their kind and were not achieved using HT strategies, but rather with large investment in time and effort. However, understanding how an organism's ecology shapes the function of its OR repertoire requires functional characterization of entire repertoires from a diversity of species from different orders and ecologies.

3.3. Future perspective in the functional characterization of insect ORs

The large number of ORs expressed by insect species, the large number of potential ligands, and the difficulties to functionally express ORs, have made large-scale OR deorphanization difficult. As discussed above, this is however required to elucidate the basic principles of odor coding in insects and develop odor-based control strategies. While *in vivo* approaches remain difficult and time consuming, *in vitro* approaches offer the possibility of developing automatization for HT functional characterization and screening. In addition, binding site modeling or *in silico* prediction of candidate ligands could further reduce the number of ligands to screen.

3.3.1 Automatization for HT screens

Modern drug discovery has been one of the major drivers of the development of automatization for HT screens, especially for membrane proteins such as GPCRs. Such automatization is currently being adapted to insect ORs in several laboratories. The development of sophisticated cellular screening systems using fluorescent or luminescent assays to detect calcium signaling with user-changeable 96-, 384-, and even 1536-well plates, together with vast chemical libraries, will dramatically accelerate the ability to screen libraries of candidate ligands against ORs expressed in HEK293.

Other screening stations have been developed to perform semiautomated (opusXpress[136]) and fully automated two-electrode voltage clamp on *Xenopus* oocytes, such as the Roboocyte (jointly developed by Multichannel Systems and Bayer Technology Services,[137,138] and the Hiclamp robots (developed in collaboration by HiQScreen and Multichannel Systems), which represent all-in-one solutions for HT functional screening of ligands. In combination, the Roboinject allows automated injection of RNA, DNA, or other samples into the oocytes. All these technologies are promising not only for the *de novo* deorphanization of insect ORs but also for the identification of agonists, antagonists, superagonists, and other interacting compounds. For instance, several studies on insect ORs have been published based on the use of the opusXpress system.[139,140]

3.3.2 In silico *HT screens*

Three-dimensional structures of OR proteins have not yet been determined and modeling of protein–odor interactions enabling sophisticated *in silico* screening methods are not yet possible, except for vertebrate ORs.[141] However, a recent innovative approach based on the analysis of odorant structures set up a chemical-informatics platform for OR functional studies.[142] Using a computational approach, the authors identified shared structural features from known ligands of individual *Drosophila* ORs, which are presumably required for binding to a receptor. These features were further used to screen new candidate ligands from >240,000 potential volatiles *in silico*. Functional experiments supported a high success rate of new activators and inhibitors identified by the modeling process.

This cheminformatics pipeline has been further used to successfully predict new ligands for *A. gambiae* chemosensory receptors tuned to behaviorally active odorants, leading to the exciting identification of new mosquito repellents[143,144] and attractants (to lure mosquitoes to traps).[144] To use such an approach, it is necessary to have information on functional ligands for the targeted receptor. Thus, these *in silico* analyses are currently restricted to Diptera (*Drosophila* and *Anopheles*). However, the progress made in OR functional characterization in other insect orders should provide the necessary information for the extension of this approach to other pest insects.

4. CONCLUSION

In the coming years, the combination of NGS, HT screens, and informatics will lead to the identification of a large number of insect OR-ligand(s) couples in a wide diversity of species with different ecologies. The possible development of methods inspired from what has been developed for the study of vertebrate ORs, such as the use of virus-driven expression *in vivo*,[145] may also offer interesting perspectives for insect OR studies.[15] Regardless, deorphanization of ORs from a diversity of insect species using the current available methodologies may first enable testing of whether the hypothesis of combinatorial coding of odors is a general rule among insects or not. For instance, *in silico* predictions suggest that a number of odorants at low concentrations may be detected by only a few receptors,[142] thus remodeling the upper evoked hypothesis. Second, this will allow understanding how the odor space is used/shared by different species to adapt to new environments. Last but not least, advances in HT identification and characterization of ORs in insects will lead to the identification of ORs involved in key behaviors (such as reproduction, ovipositioning, and host selection) of harmful insects offering the opportunity to identify new generations of attractants/repellents that will aid the control of crop pests and deadly insect-borne diseases worldwide.

REFERENCES

1. De Bruyne M, Baker TC. Odor detection in insects: volatile codes. *J Chem Ecol.* 2008;34:882–897.
2. Schneider D. Insect antennae. *Annu Rev Entomol.* 1964;9:103–122.
3. de Bruyne M, Foster K, Carlson JR. Odor coding in the *Drosophila* antenna. *Neuron.* 2001;30(2):537–552.
4. Binyameen M, Anderson P, Ignell R, Seada MA, Hansson BS, Schlyter F. Spatial organization of antennal olfactory sensory neurons in the female *Spodoptera littoralis* moth: differences in sensitivity and temporal characteristics. *Chem Senses.* 2012;37(7): 613–629.
5. Coutinho-Abreu IV, McInally S, Forster L, Luck R, Ray A. Odor coding in a disease-transmitting herbivorous insect, the Asian citrus psyllid. *Chem Senses.* 2014;39(6): 539–549.
6. Leal WS. Odorant reception in insects: roles of receptors, binding proteins, and degrading enzymes. *Annu Rev Entomol.* 2013;58:373–391.

7. Ishida Y, Leal WS. Rapid inactivation of a moth pheromone. *Proc Natl Acad Sci USA*. 2005;102(39):14075–14079.
8. Durand N, Carot-Sans G, Bozzolan F, et al. Degradation of pheromone and plant volatile components by a same odorant-degrading enzyme in the cotton leafworm, *Spodoptera littoralis*. *PLoS One*. 2011;6(12):e29147.
9. Chertemps T, Francois A, Durand N, et al. A carboxylesterase, Esterase-6, modulates sensory physiological and behavioral response dynamics to pheromone in *Drosophila*. *BMC Biol*. 2012;10:56.
10. Benton R, Vannice KS, Gomez-Diaz C, Vosshall LB. Variant ionotropic glutamate receptors as chemosensory receptors in *Drosophila*. *Cell*. 2009;136(1):149–162.
11. Croset V, Rytz R, Cummins SF, et al. Ancient protostome origin of chemosensory ionotropic glutamate receptors and the evolution of insect taste and olfaction. *PLoS Genet*. 2010;6(8):e1001064.
12. Silbering AF, Rytz R, Grosjean Y, et al. Complementary function and integrated wiring of the evolutionarily distinct *Drosophila* olfactory subsystems. *J Neurosci*. 2011;31(38):13357–13375.
13. Robertson HM, Warr CG, Carlson JR. Molecular evolution of the insect chemoreceptor gene superfamily in *Drosophila melanogaster*. *Proc Natl Acad Sci USA*. 2003;100 (Suppl. 2):14537–14542.
14. Missbach C, Dweck HK, Vogel H, et al. Evolution of insect olfactory receptors. *Elife (Cambridge)*. 2014;3:e02115.
15. Sakurai T, Nakagawa T, Mitsuno H, et al. Identification and functional characterization of a sex pheromone receptor in the silkmoth *Bombyx mori*. *Proc Natl Acad Sci USA*. 2004;101(47):16653–16658.
16. Hallem EA, Carlson JR. Coding of odors by a receptor repertoire. *Cell*. 2006;125(1): 143–160.
17. Tanaka K, Uda Y, Ono Y, et al. Highly selective tuning of a silkworm olfactory receptor to a key mulberry leaf volatile. *Curr Biol*. 2009;19(11):881–890.
18. Carey AF, Wang G, Su CY, Zwiebel LJ, Carlson JR. Odorant reception in the malaria mosquito *Anopheles gambiae*. *Nature*. 2010;464(7285):66–71.
19. Larsson MC, Domingos AI, Jones WD, Chiappe ME, Amrein H, Vosshall LB. Or83b encodes a broadly expressed odorant receptor essential for *Drosophila* olfaction. *Neuron*. 2004;43(5):703–714.
20. Benton R, Sachse S, Michnick SW, Vosshall LB. Atypical membrane topology and heteromeric function of *Drosophila* odorant receptors in vivo. *PLoS Biol*. 2006;4(2):e20.
21. Smart R, Kiely A, Beale M, et al. *Drosophila* odorant receptors are novel seven transmembrane domain proteins that can signal independently of heterotrimeric G proteins. *Insect Biochem Mol Biol*. 2008;38(8):770–780.
22. Sato K, Pellegrino M, Nakagawa T, Vosshall LB, Touhara K. Insect olfactory receptors are heteromeric ligand-gated ion channels. *Nature*. 2008;452(7190):1002–1006.
23. Wicher D, Schafer R, Bauernfeind R, et al. *Drosophila* odorant receptors are both ligand-gated and cyclic-nucleotide-activated cation channels. *Nature*. 2008;452(7190): 1007–1011.
24. Kirkness EF, Haas BJ, Sun W, et al. Genome sequences of the human body louse and its primary endosymbiont provide insights into the permanent parasitic lifestyle. *Proc Natl Acad Sci USA*. 2010;107(27):12168–12173.
25. Zhou X, Slone JD, Rokas A, et al. Phylogenetic and transcriptomic analysis of chemosensory receptors in a pair of divergent ant species reveals sex-specific signatures of odor coding. *PLoS Genet*. 2012;8(8):e1002930.
26. Couto A, Alenius M, Dickson BJ. Molecular, anatomical, and functional organization of the *Drosophila* olfactory system. *Curr Biol*. 2005;15(17):1535–1547.

27. Buck L, Axel R. A novel multigene family may encode odorant receptors: a molecular basis for odor recognition. *Cell.* 1991;65:175–187.
28. Mombaerts P. Love at first smell—the 2004 Nobel Prize in Physiology or Medicine. *N Engl J Med.* 2004;351(25):2579–2580.
29. Ben-Arie N, Lancet D, Taylor C, et al. Olfactory receptor gene cluster on human chromosome 17: possible duplication of an ancestral receptor repertoire. *Hum Mol Genet.* 1994;3(2):229–235.
30. Ngai J, Chess A, Dowling MM, Necles N, Macagno ER, Axel R. Coding of olfactory information: topography of odorant receptor expression in the catfish olfactory epithelium. *Cell.* 1993;72(5):667–680.
31. Nef S, Allaman I, Fiumelli H, De Castro E, Nef P. Olfaction in birds: differential embryonic expression of nine putative odorant receptor genes in the avian olfactory system. *Mech Dev.* 1996;55(1):65–77.
32. Mombaerts P. Molecular biology of odorant receptors in vertebrates. *Annu Rev Neurosci.* 1999;22:487–509.
33. Danty E, Cornuet J-M, Masson C. Honeybees have putative olfactory receptor proteins similar to those of vertebrates. *C R Acad Sci Paris.* 1994;317:1073–1079.
34. Adams MD, Celniker SE, Holt RA, et al. The genome sequence of *Drosophila melanogaster. Science.* 2000;287(5461):2185–2195.
35. Sanger F, Nicklen S, Coulson AR. DNA sequencing with chain-terminating inhibitors. *Proc Natl Acad Sci USA.* 1977;74(12):5463–5467.
36. Holt RA, Subramanian GM, Halpern A, et al. The genome sequence of the malaria mosquito *Anopheles gambiae. Science.* 2002;298(5591):129–149.
37. Mita K, Kasahara M, Sasaki S, et al. The genome sequence of silkworm, *Bombyx mori. DNA Res.* 2004;11(1):27–35.
38. Xia Q, Zhou Z, Lu C, et al. A draft sequence for the genome of the domesticated silkworm (*Bombyx mori*). *Science.* 2004;306(5703):1937–1940.
39. T.H.G.S. Consortium. Insights into social insects from the genome of the honeybee *Apis mellifera. Nature.* 2006;443(7114):931–949.
40. Richards S, Gibbs RA, Weinstock GM, et al. The genome of the model beetle and pest *Tribolium castaneum. Nature.* 2008;452(7190):949–955.
41. Buermans HPJ, den Dunnen JT. Next generation sequencing technology: advances and applications. *Biochim Biophys Acta.* 2014;1842(10):1932–1941.
42. Zhan S, Merlin C, Boore JL, Reppert SM. The monarch butterfly genome yields insights into long-distance migration. *Cell.* 2011;147(5):1171–1185.
43. The Heliconius Genome Consortium. Islands of divergence underlie adaptive radiation in a butterfly genome. *Nature.* 2012;487:94–98.
44. You M, Yue Z, He W, et al. A heterozygous moth genome provides insights into herbivory and detoxification. *Nat Genet.* 2013;45(2):220–225.
45. Giardine B, Riemer C, Hardison RC, et al. Galaxy: a platform for interactive large-scale genome analysis. *Genome Res.* 2005;15(10):1451–1455.
46. Blankenberg D, Hillman-Jackson J. Analysis of next-generation sequencing data using galaxy. *Methods Mol Biol.* 2014;1150:21–43.
47. Clyne PJ, Warr CG, Freeman MR, Lessing D, Kim J, Carlson JR. A novel family of divergent seven-transmembrane proteins: candidate odorant receptors in Drosophila. *Neuron.* 1999;22(2):327–338.
48. Gao Q, Chess A. Identification of candidate *Drosophila* olfactory receptors from genomic DNA sequence. *Genomics.* 1999;60(1):31–39.
49. Vosshall LB, Amrein H, Morozov PS, Rzhetsky A, Axel R. A spatial map of olfactory receptor expression in the *Drosophila* antenna. *Cell.* 1999;96(5):725–736.
50. Krieger J, Klink O, Mohl C, Raming K, Breer H. A candidate olfactory receptor subtype highly conserved across different insect orders. *J Comp Physiol A.* 2003;189(7):519–526.

51. Melo AC, Rutzler M, Pitts RJ, Zwiebel LJ. Identification of a chemosensory receptor from the yellow fever mosquito, Aedes aegypti, that is highly conserved and expressed in olfactory and gustatory organs. *Chem Senses*. 2004;29(5):403–410.
52. Malpel S, Merlin C, Francois MC, Jacquin-Joly E. Molecular identification and characterization of two new Lepidoptera chemoreceptors belonging to the *Drosophila melanogaster* OR83b family. *Insect Mol Biol*. 2008;17(5):587–596.
53. Fox AN, Pitts RJ, Robertson HM, Carlson JR, Zwiebel LJ. Candidate odorant receptors from the malaria vector mosquito *Anopheles gambiae* and evidence of down-regulation in response to blood feeding. *Proc Natl Acad Sci USA*. 2001;98(25): 14693–14697.
54. Hill CA, Fox AN, Pitts RJ, et al. G protein-coupled receptors in Anopheles gambiae. *Science*. 2002;298(5591):176–178.
55. Kim J, Moriyama EN, Warr CG, Clyne PJ, Carlson JR. Identification of novel multi-transmembrane proteins from genomic databases using quasi-periodic structural properties. *Bioinformatics*. 2000;16(9):767–775.
56. Engsontia P, Sanderson AP, Cobb M, Walden KK, Robertson HM, Brown S. The red flour beetle's large nose: an expanded odorant receptor gene family in Tribolium castaneum. *Insect Biochem Mol Biol*. 2008;38(4):387–397.
57. Bohbot J, Pitts RJ, Kwon HW, Rutzler M, Robertson HM, Zwiebel LJ. Molecular characterization of the *Aedes aegypti* odorant receptor gene family. *Insect Mol Biol*. 2007;16(5):525–537.
58. Marinotti O, Cerqueira GC, de Almeida LG, et al. The genome of *Anopheles darlingi*, the main neotropical malaria vector. *Nucleic Acids Res*. 2013;41(15):7387–7400.
59. Arensburger P, Megy K, Waterhouse RM, et al. Sequencing of *Culex quinquefasciatus* establishes a platform for mosquito comparative genomics. *Science*. 2010;330(6000): 86–88.
60. Nozawa M, Nei M. Evolutionary dynamics of olfactory receptor genes in *Drosophila* species. *Proc Natl Acad Sci USA*. 2007;104(17):7122–7127.
61. Obiero GF, Mireji PO, Nyanjom SR, Christoffels A, Robertson HM, Masiga DK. Odorant and gustatory receptors in the tsetse fly *Glossina morsitans morsitans*. *PLoS Negl Trop Dis*. 2014;8(4):e2663.
62. Andersson MN, Videvall E, Walden KK, Harris MO, Robertson HM, Lofstedt C. Sex- and tissue-specific profiles of chemosensory gene expression in a herbivorous gall-inducing fly (Diptera: Cecidomyiidae). *BMC Genomics*. 2014;15(1):501.
63. Cao D, Liu Y, Walker WB, Li J, Wang G. Molecular characterization of the *Aphis gossypii* olfactory receptor gene families. *PLoS One*. 2014;9(6):e101187.
64. Nygaard S, Zhang G, Schiott M, et al. The genome of the leaf-cutting ant *Acromyrmex echinatior* suggests key adaptations to advanced social life and fungus farming. *Genome Res*. 2011;21(8):1339–1348.
65. Robertson HM, Wanner KW. The chemoreceptor superfamily in the honey bee, *Apis mellifera*: expansion of the odorant, but not gustatory, receptor family. *Genome Res*. 2006;16(11):1395–1403.
66. Koch SI, Groh K, Vogel H, Hansson BS, Kleineidam CJ, Grosse-Wilde E. Caste-specific expression patterns of immune response and chemosensory related genes in the leaf-cutting ant, *Atta vollenweideri*. *PLoS One*. 2013;8(11):e81518.
67. Smith CD, Zimin A, Holt C, et al. Draft genome of the globally widespread and invasive Argentine ant (*Linepithema humile*). *Proc Natl Acad Sci USA*. 2011;108(14): 5673–5678.
68. Robertson HM, Gadau J, Wanner KW. The insect chemoreceptor superfamily of the parasitoid jewel wasp *Nasonia vitripennis*. *Insect Mol Biol*. 2010;19(Suppl. 1):121–136.
69. Smith CR, Smith CD, Robertson HM, et al. Draft genome of the red harvester ant *Pogonomyrmex barbatus*. *Proc Natl Acad Sci USA*. 2011;108(14):5667–5672.

70. Wurm Y, Wang J, Riba-Grognuz O, et al. The genome of the fire ant *Solenopsis invicta*. *Proc Natl Acad Sci USA*. 2011;108(14):5679–5684.
71. Engsontia P, Sangket U, Chotigeat W, Satasook C. Molecular evolution of the odorant and gustatory receptor genes in lepidopteran insects: implications for their adaptation and speciation. *J Mol Evol*. 2014;79(1-2):21–39.
72. Robinson GE, Hackett KJ, Purcell-Miramontes M, et al. Creating a buzz about insect genomes. *Science*. 2011;331(6023):1386.
73. i5K Consortium. The i5K Initiative: advancing arthropod genomics for knowledge, human health, agriculture, and the environment. *J Hered*. 2013;104(5):595–600.
74. Wanner KW, Anderson AR, Trowell SC, Theilmann DA, Robertson HM, Newcomb RD. Female-biased expression of odourant receptor genes in the adult antennae of the silkworm, *Bombyx mori*. *Insect Mol Biol*. 2007;16(1):107–119.
75. Robertson HM, Martos R, Sears CR, Todres EZ, Walden KK, Nardi JB. Diversity of odourant binding proteins revealed by an expressed sequence tag project on male *Manduca sexta* moth antennae. *Insect Mol Biol*. 1999;8:501–518.
76. Whitfield CW, Band MR, Bonaldo MF, et al. Annotated expressed sequence tags and cDNA microarrays for studies of brain and behavior in the honey bee. *Genome Res*. 2002;12(4):555–566.
77. Jordan MD, Stanley D, Marshall SD, et al. Expressed sequence tags and proteomics of antennae from the tortricid moth, *Epiphyas postvittana*. *Insect Mol Biol*. 2008;17(4):361–373.
78. Patch HM, Velarde RA, Walden KK, Robertson HM. A candidate pheromone receptor and two odorant receptors of the hawkmoth *Manduca sexta*. *Chem Senses*. 2009;34(4):305–316.
79. Legeai F, Malpel S, Montagne N, et al. An Expressed Sequence Tag collection from the male antennae of the Noctuid moth *Spodoptera littoralis*: a resource for olfactory and pheromone detection research. *BMC Genomics*. 2011;12(1):86.
80. Jacquin-Joly E, Legeai F, Montagné N, et al. Candidate chemosensory genes in female antennae of the noctuid moth *Spodoptera littoralis*. *Int J Biol Sci*. 2012;8:1036.
81. Mamidala P, Wijeratne AJ, Wijeratne S, et al. Identification of odor-processing genes in the emerald ash borer, *Agrilus planipennis*. *PLoS One*. 2013;8(2):e56555.
82. Wang J, Li DZ, Min SF, Mi F, Zhou SS, Wang MQ. Analysis of chemosensory gene families in the beetle *Monochamus alternatus* and its parasitoid *Dastarcus helophoroides*. *Comp Biochem Physiol Part D Genomics Proteomics*. 2014;11C:1–8.
83. Andersson MN, Grosse-Wilde E, Keeling CI, et al. Antennal transcriptome analysis of the chemosensory gene families in the tree killing bark beetles, *Ips typographus* and *Dendroctonus ponderosae* (Coleoptera: Curculionidae: Scolytinae). *BMC Genomics*. 2013;14:198.
84. Mitchell RF, Hughes DT, Luetje CW, et al. Sequencing and characterizing odorant receptors of the cerambycid beetle *Megacyllene caryae*. *Insect Biochem Mol Biol*. 2012;42(7):499–505.
85. Wang ZL, Liu TT, Huang ZY, Wu XB, Yan WY, Zeng ZJ. Transcriptome analysis of the Asian honey bee *Apis cerana cerana*. *PLoS One*. 2012;7(10):e47954.
86. Gress JC, Robertson HM, Weaver DK, Dlakic M, Wanner KW. Odorant receptors of a primitive hymenopteran pest, the wheat stem sawfly. *Insect Mol Biol*. 2013;22(6):659–667.
87. Nishimura O, Brillada C, Yazawa S, Maffei ME, Arimura G. Transcriptome pyrosequencing of the parasitoid wasp *Cotesia vestalis*: genes involved in the antennal odorant-sensory system. *PLoS One*. 2012;7(11):e50664.
88. Gu SH, Sun L, Yang RN, et al. Molecular characterization and differential expression of olfactory genes in the antennae of the black cutworm moth *Agrotis ipsilon*. *PLoS One*. 2014;9(8):e103420.

89. Cao D, Liu Y, Wei J, et al. Identification of candidate olfactory genes in *Chilo suppressalis* by antennal transcriptome analysis. *Int J Biol Sci*. 2014;10(8):846–860.
90. Bengtsson JM, Trona F, Montagné N, et al. Putative chemosensory receptors of the codling moth, *Cydia pomonella*, identified by antennal transcriptome analysis. *PLoS One*. 2012;7(2):e31620.
91. Zhang S, Zhang Z, Wang H, Kong X. Antennal transcriptome analysis and comparison of olfactory genes in two sympatric defoliators, *Dendrolimus houi* and *Dendrolimus kikuchii* (Lepidoptera: Lasiocampidae). *Insect Biochem Mol Biol*. 2014;52C:69–81.
92. Liu Y, Gu S, Zhang Y, Guo Y, Wang G. Candidate olfaction genes identified within the *Helicoverpa armigera* Antennal Transcriptome. *PLoS One*. 2012;7(10):e48260.
93. Liu NY, Xu W, Papanicolaou A, Dong SL, Anderson A. Identification and characterization of three chemosensory receptor families in the cotton bollworm *Helicoverpa armigera*. *BMC Genomics*. 2014;15(1):597.
94. Grosse-Wilde E, Kuebler LS, Bucks S, Vogel H, Wicher D, Hansson BS. Antennal transcriptome of *Manduca sexta*. *Proc Natl Acad Sci USA*. 2011;108(18):7449–7454.
95. Howlett N, Dauber KL, Shukla A, et al. Identification of chemosensory receptor genes in *Manduca sexta* and knockdown by RNA interference. *BMC Genomics*. 2012;13:211.
96. Zhang YN, Jin JY, Jin R, et al. Differential expression patterns in chemosensory and non-chemosensory tissues of putative chemosensory genes identified by transcriptome analysis of insect pest the purple stem borer *Sesamia inferens* (Walker). *PLoS One*. 2013;8(7):e69715.
97. Glaser N, Gallot A, Legeai F, et al. Candidate chemosensory genes in the Stemborer *Sesamia nonagrioides*. *Int J Biol Sci*. 2013;9:481–495.
98. Poivet E, Gallot A, Montagné N, Glaser N, Legeai F, Jacquin-Joly E. A comparison of the olfactory gene repertoires of adults and larvae in the noctuid moth *Spodoptera littoralis*. *PLoS One*. 2013;8(4):e60263.
99. Li ZQ, Zhang S, Ma Y, et al. First transcriptome and digital gene expression analysis in Neuroptera with an emphasis on chemoreception genes in *Chrysopa pallens* (Rambur). *PLoS One*. 2013;8(6):e67151.
100. Kreher SA, Kwon JY, Carlson JR. The molecular basis of odor coding in the *Drosophila* larva. *Neuron*. 2005;46(3):445–456.
101. Mathew D, Martelli C, Kelley-Swift E, et al. Functional diversity among sensory receptors in a *Drosophila* olfactory circuit. *Proc Natl Acad Sci USA*. 2013;110(23):E2134–E2143.
102. Wang G, Carey AF, Carlson JR, Zwiebel LJ. Molecular basis of odor coding in the malaria vector mosquito *Anopheles gambiae*. *Proc Natl Acad Sci USA*. 2010;107(9): 4418–4423.
103. Wetzel CH, Behrendt HJ, Gisselmann G, Stortkuhl KF, Hovemann B, Hatt H. Functional expression and characterization of a *Drosophila* odorant receptor in a heterologous cell system. *Proc Natl Acad Sci USA*. 2001;98(16):9377–9380.
104. Nakagawa T, Sakurai T, Nishioka T, Touhara K. Insect sex-pheromone signals mediated by specific combinations of olfactory receptors. *Science*. 2005;307(5715): 1638–1642.
105. Thomas P, Smart TG. HEK293 cell line: a vehicle for the expression of recombinant proteins. *J Pharmacol Toxicol Methods*. 2005;51(3):187–200.
106. Schneider EH, Seifert R. Sf9 cells: a versatile model system to investigate the pharmacological properties of G protein-coupled receptors. *Pharmacol Ther*. 2010;128(3): 387–418.
107. Grosse-Wilde E, Svatos A, Krieger J. A pheromone-binding protein mediates the bombykol-induced activation of a pheromone receptor in vitro. *Chem Senses*. 2006;31:547–555.

108. Bohbot JD, Jones PL, Wang G, Pitts RJ, Pask GM, Zwiebel LJ. Conservation of indole responsive odorant receptors in mosquitoes reveals an ancient olfactory trait. *Chem Senses*. 2011;36(2):149–160.
109. Jones PL, Pask GM, Rinker DC, Zwiebel LJ. Functional agonism of insect odorant receptor ion channels. *Proc Natl Acad Sci USA*. 2011;108(21):8821–8825.
110. Pregitzer P, Schubert M, Breer H, Hansson BS, Sachse S, Krieger J. Plant odorants interfere with detection of sex pheromone signals by male *Heliothis virescens*. *Front Cell Neurosci*. 2012;6:42.
111. Forstner M, Breer H, Krieger J. A receptor and binding protein interplay in the detection of a distinct pheromone component in the silkmoth *Antheraea polyphemus*. *Int J Biol Sci*. 2009;5(7):745–757.
112. Corcoran JA, Jordan MD, Carraher C, Newcomb RD. A novel method to study insect olfactory receptor function using HEK293 cells. *Insect Biochem Mol Biol*. 2014;54C: 22–32.
113. Kumar BN, Taylor RW, Pask GM, Zwiebel LJ, Newcomb RD, Christie DL. A conserved aspartic acid is important for agonist (VUAA1) and odorant/tuning receptor-dependent activation of the insect odorant co-receptor (Orco). *PLoS One*. 2013;8(7):e70218.
114. Kiely A, Authier A, Kralicek AV, Warr CG, Newcomb RD. Functional analysis of a *Drosophila melanogaster* olfactory receptor expressed in Sf9 cells. *J Neurosci Methods*. 2007;159(2):189–194.
115. Jordan MD, Anderson A, Begum D, et al. Odorant receptors from the light brown apple moth (*Epiphyas postvittana*) recognize important volatile compounds produced by plants. *Chem Senses*. 2009;34(5):383–394.
116. Anderson AR, Wanner KW, Trowell SC, et al. Molecular basis of female-specific odorant responses in *Bombyx mori*. *Insect Biochem Mol Biol*. 2009;39(3):189–197.
117. Claudianos C, Lim J, Young M, et al. Odor memories regulate olfactory receptor expression in the sensory periphery. *Eur J Neurosci*. 2014;39(10):1642–1654.
118. Lundin C, Kall L, Kreher SA, et al. Membrane topology of the *Drosophila* OR83b odorant receptor. *FEBS Lett*. 2007;581(29):5601–5604.
119. Tsitoura P, Andronopoulou E, Tsikou D, et al. Expression and membrane topology of *Anopheles gambiae* odorant receptors in lepidopteron insect cells. *PLoS One*. 2010;5(11): e15428.
120. German PF, van der Poel S, Carraher C, Kralicek AV, Newcomb RD. Insights into subunit interactions within the insect olfactory receptor complex using FRET. *Insect Biochem Mol Biol*. 2013;43(2):138–145.
121. Hamada S, Tabuchi M, Toyota T, et al. Giant vesicles functionally expressing membrane receptors for an insect pheromone. *Chem Commun (Camb)*. 2014;50(22): 2958–2961.
122. Carraher C, Nazmi AR, Newcomb RD, Kralicek A. Recombinant expression, detergent solubilisation and purification of insect odorant receptor subunits. *Protein Expr Purif*. 2013;90(2):160–169.
123. Dobritsa AA, van der Goes van Naters W, Warr CG, Steinbrecht RA, Carlson JR. Integrating the molecular and cellular basis of odor coding in the *Drosophila* antenna. *Neuron*. 2003;37(5):827–841.
124. Syed Z, Ishida Y, Taylor K, Kimbrell DA, Leal WS. Pheromone reception in fruit flies expressing a moth's odorant receptor. *Proc Natl Acad Sci USA*. 2006;103(44): 16538–16543.
125. Montagné N, Chertemps T, Brigaud I, et al. Functional characterization of a sex pheromone receptor in the pest moth *Spodoptera littoralis* by heterologous expression in *Drosophila*. *Eur J Neurosci*. 2012;36(5):2588–2596.

126. Ronderos DS, Lin CC, Potter CJ, Smith DP. Farnesol-detecting olfactory neurons in *Drosophila*. *J Neurosci*. 2014;34(11):3959–3968.
127. Kurtovic A, Widmer A, Dickson BJ. A single class of olfactory neurons mediates behavioural responses to a *Drosophila* sex pheromone. *Nature*. 2007;446(7135):542–546.
128. Benton R, Vannice KS, Vosshall LB. An essential role for a CD36-related receptor in pheromone detection in *Drosophila*. *Nature*. 2007;450(7167):289–293.
129. Jin X, Ha TS, Smith DP. SNMP is a signaling component required for pheromone sensitivity in *Drosophila*. *Proc Natl Acad Sci USA*. 2008;105(31):10996–11001.
130. Sakurai T, Mitsuno H, Haupt SS, et al. A single sex pheromone receptor determines chemical response specificity of sexual behavior in the silkmoth *Bombyx mori*. *PLoS Genet*. 2011;7(6):e1002115.
131. Stortkuhl KF, Kettler R. Functional analysis of an olfactory receptor in *Drosophila melanogaster*. *Proc Natl Acad Sci USA*. 2001;98(16):9381–9385.
132. Ueira-Vieira C, Kimbrell DA, de Carvalho WJ, Leal WS. Facile functional analysis of insect odorant receptors expressed in the fruit fly: validation with receptors from taxonomically distant and closely related species. *Cell Mol Life Sci*. 2014;71(23):4675–4680.
133. Hiroi M, Tanimura T, Marion-Poll F. Hedonic taste in *Drosophila* revealed by olfactory receptors expressed in taste neurons. *PLoS One*. 2008;3(7):e2610.
134. Hallem EA, Carlson JR. The spatial code for odors is changed by conditioning. *Neuron*. 2004;42(3):359–361.
135. Kreher SA, Mathew D, Kim J, Carlson JR. Translation of sensory input into behavioral output via an olfactory system. *Neuron*. 2008;59(1):110–124.
136. Papke RL, Stokes C. Working with OpusXpress: methods for high volume oocyte experiments. *Methods*. 2010;51(1):121–133.
137. Schnizler K, Kuster M, Methfessel C, Fejtl M. The roboocyte: automated cDNA/mRNA injection and subsequent TEVC recording on Xenopus oocytes in 96-well microtiter plates. *Receptors Channels*. 2003;9(1):41–48.
138. Leisgen C, Kuester M, Methfessel C. The roboocyte: automated electrophysiology based on *Xenopus* oocytes. *Methods Mol Biol*. 2007;403:87–109.
139. Chen S, Luetje CW. Identification of new agonists and antagonists of the insect odorant receptor co-receptor subunit. *PLoS One*. 2012;7(5):e36784.
140. Chen S, Luetje CW. Trace amines inhibit insect odorant receptor function through antagonism of the co-receptor subunit. *F1000Res*. 2014;3:84.
141. Triballeau N, Van Name E, Laslier G, et al. High-potency olfactory receptor agonists discovered by virtual high-throughput screening: molecular probes for receptor structure and olfactory function. *Neuron*. 2008;60(5):767–774.
142. Boyle SM, McInally S, Ray A. Expanding the olfactory code by in silico decoding of odor-receptor chemical space. *Elife*. 2013;2:e01120.
143. Kain P, Boyle SM, Tharadra SK, et al. Odour receptors and neurons for DEET and new insect repellents. *Nature*. 2013;502(7472):507–512.
144. Tauxe GM, MacWilliam D, Boyle SM, Guda T, Ray A. Targeting a dual detector of skin and CO2 to modify mosquito host seeking. *Cell*. 2013;155(6):1365–1379.
145. Touhara K, Sengoku S, Inaki K, et al. Functional identification and reconstitution of an odorant receptor in single olfactory neurons. *Proc Natl Acad Sci USA*. 1999;96:4040–4045.
146. Smadja C, Shi P, Butlin RK, Robertson HM. Large gene family expansions and adaptive evolution for odorant and gustatory receptors in the pea aphid, *Acyrthosiphon pisum*. *Mol Biol Evol*. 2009;26(9):2073–2086.

CHAPTER FOUR

Olfactory Disruption: Toward Controlling Important Insect Vectors of Disease

Jackson T. Sparks, Jonathan D. Bohbot, Joseph C. Dickens[1]
United States Department of Agriculture, Beltsville Agricultural Research Center, Invasive Insect Biocontrol and Behavior Laboratory, Beltsville, Maryland, USA
[1]Corresponding author: e-mail address: joseph.dickens@ars.usda.gov

Contents

Abstract

Chemical repellents are used to decrease contacts between insect disease vectors and their hosts, thus reducing the probability of disease transmission. The molecular mechanisms by which repellents have their effects are poorly understood and remain a controversial topic. Here, we present recent results of studies aimed at a more thorough understanding of the mode of action of repellents and discuss the implications of these findings for future research and development of novel or improved repellents.

Progress in Molecular Biology and Translational Science, Volume 130
ISSN 1877-1173
http://dx.doi.org/10.1016/bs.pmbts.2014.11.004

1. INTRODUCTION

Insect vectors of human diseases burden health systems and cause millions of deaths yearly, particularly in developing countries in South and Central America, and Asia.[1–3] Millions of dollars are spent each year on research and production of repellent compounds to prevent insect bites that lead to disease transmission.[4] The insect vectors responsible for the spread of human diseases include Diptera (mosquitoes and biting flies), Hemiptera (true bugs), Anoplura (lice), and Siphonaptera (fleas).[5–7] The sense organs of these insects are specifically tuned to detect chemical and physical signals emitted by potential hosts in order to locate a blood meal. Without eradicating the viruses and other zoonotic entities responsible for the diseases, prevention of disease transmission depends on the disruption of host seeking by the insect vector. While insect repellents have been used for at least 80 years to decrease contacts between insect vectors and their vertebrate hosts,[8] safer alternative compounds with greater efficacy and duration of activity are needed.

In this review, we explore the olfactory system of insects with emphasis on those that vector diseases. While much of the work highlighted here is from mosquitoes, the molecular features discussed are relevant to all biting insects. The chemosensory organs, neural pathways, cells, and genes involved in the detection of chemical cues are described to provide context for understanding how repellents function. Then, strategies aimed at discovery of the next generation of repellents are discussed.

2. DETECTION OF OLFACTORY SIGNALS BY INSECT VECTORS

Chemosensory genes involved in detection and transduction of olfactory signals in insects represent molecular targets for repellents. These proteins influence olfactory-mediated behaviors like host seeking and include: (1) odorant-binding proteins (OBPs), (2) chemosensory proteins (CSPs), (3) ionotropic receptors (IRs), (4) odorant receptors (ORs), (5) gustatory receptors (GRs), and (6) sensory neuron membrane proteins (SNMPs) (Fig. 1). The genomes of 27 vector species have been or are currently being sequenced (VectorBase; http://www.vectorbase.org) and chemosensory genes of 21 mosquito species (1 *Aedes*, 19 *Anopheles*, and 1 *Culex* species) and the savannah tsetse fly *Glossina morsitans* have been annotated.[9–17]

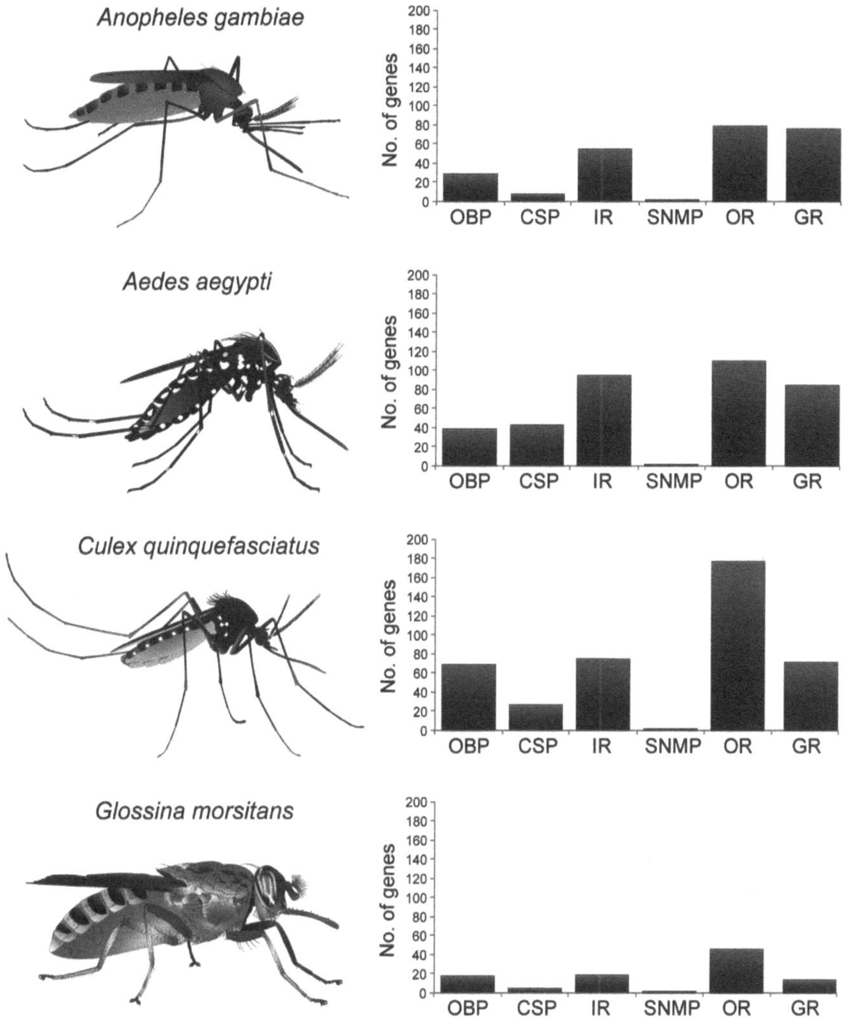

Figure 1 Chemosensory genes of disease-transmitting dipterans. With the availability of whole-genome sequences from mosquitoes (*Anopheles gambiae*, *Aedes aegypti*, and *Culex quinquefasciatus*) and the savannah tsetse fly (*Glossina morsitans*), the odorant-binding protein (OBP), chemosensory protein (CSP), ionotropic receptor (IR), sensory neuron membrane protein (SNMP), odorant receptor (OR), and gustatory receptor (GR) gene families have been identified. A comparison of the chemosensory gene profile of each species suggests *G. morsitans* can discriminate fewer numbers of odors than mosquitoes. Homologous chemosensory genes from these species may represent common chemosensory sensitivities, e.g., attractive host odors.

A cursory comparison of the number of olfactory genes between these species suggests that mosquitoes have a more elaborate sense of smell than the fly, *G. morsitans* (Fig. 1).

Sensilla, cuticular structures on the body of the insect, are comprised of a variety of cells, biochemical pathways, and distinct protein families that function in the detection of chemical signals. While olfactory sensilla are positioned on the surface of the antennae and maxillary palps, most gustatory sensilla are located on the mouthparts, tarsi, and wings (Fig. 2A). Chemical detection occurs within the shaft of a sensillum where dendrites of sensory neurons are immersed in its aqueous lumen (Fig. 2B). Three support cells modulate the physiological conditions responsible for protecting and preserving the function of these dendrites, including releasing soluble binding proteins into the lumen.

Based on their morphological and physiological properties, two types of chemosensory sensilla have been described. Olfactory sensilla have multiple pores and the sensory neurons housed within them detect low concentrations of airborne chemicals (Fig. 2B). Gustatory sensilla possess a terminal pore through which relatively high concentrations of chemicals pass to make contact with the dendrites of sensory neurons. Odorant detection depends on the transportation of chemical signals from pores to the membrane receptors on the surface of sensory neurons (Fig. 2C).

2.1. Odorant-binding and chemosensory proteins

OBPs[18] and CSPs[19,20] (also known as Olfactory System D or sensory appendage proteins) belong to two phylogenetically distinct insect gene families involved in chemosensation, release of semiochemicals, and development.[21,22] OBPs and CSPs are compact globular water-soluble proteins with ligand-binding capabilities and may be involved in functions other than chemosensation (Fig. 1). In sensilla housing chemosensory neurons,[23–28] these proteins are secreted by the support cells in millimolar concentrations into the sensillum lymph.[29]

OBPs and CSPs have been proposed as transporters of chemical signals from the air to ORs[18] and GRs[30] by binding odorants with various degrees of specificity using at least three separate modes of action. In the vinegar fly, *Drosophila melanogaster*, the OBP–ligand complex may elicit OR activation,[31,32] although this model has been challenged by a recent study and remains controversial.[33] In moths, OBPs rely on a pH-activated α-helix to eject the odorant from the binding pocket.[34–39] A third mechanism used

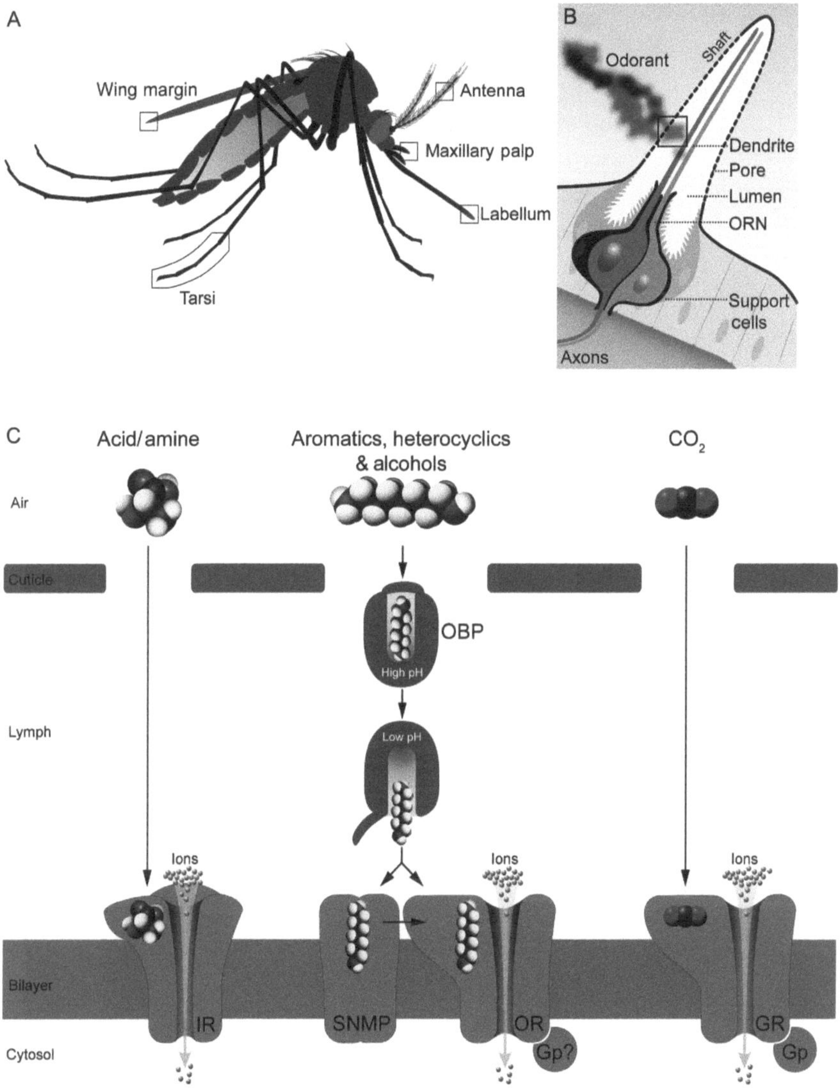

Figure 2 The peripheral olfactory system of insect vectors of disease. (A) Odorants are detected by sensory neurons housed within sensilla on the antennae, maxillary palps, labella, wing margins, and tarsi. (B) Odorants penetrate the olfactory sensillum through pores and interact with receptors on the dendritic end of olfactory receptor neurons (ORNs). (C) Different receptor families are specialized in different types of odorants. Ionotropic receptors (IR) detect acids and amines. Aromatics and other compounds are transported by odorant-binding proteins (OBPs) and may interact directly with odorant receptors (ORs) or via sensory neuron membrane proteins (SNMPs). CO_2 is detected by gustatory receptors (GRs). ORs and GRs are ligand-gated ion channels that may interact with G-proteins (GPs).

by mosquitoes to detect and release odorants[40,41] involves binding of the ligand by an OBP at high pH with the ligand being covered by a C-terminus loop that acts like a "lid" (Fig. 2C). In the proximity of the negatively charged membrane (low pH) of the sensory neuron, this cover is flipped open to release the ligand for activation of the receptor.[42]

The role of OBPs and CSPs in defining the phenotype of sensory neurons within a sensillum is unclear and remains an active topic of research. The selectivity of specific OBPs and CSPs may be narrow, but in general, OBPs accommodate a wide range of ligands in the presumed binding pocket,[39] an observation consistent with mosquito OBPs[43] and CSPs.[22] In fact, a common problem associated with both functional and structural studies of OBPs and CSPs is the presence of serendipitous ligands occupying the binding pocket.[43] OBPs are abundantly expressed in the labella and tarsi of *Aedes aegypti*,[44] suggesting a transport function for gustatory stimuli. While the role of OBPs and CSPs in mosquito gustatory responses is unknown, they are required for normal gustatory responses in other diptera.[30,45–48]

2.2. Odorant receptors

Insect *Or* genes[49] belong to a divergent gene family of seven-transmembrane proteins, which unlike mammalian ORs that are classic G-protein-coupled receptors, have an inverted membrane topology.[50] Insect ORs function as heteromers consisting of a ligand-sensing OR and an obligate coreceptor named ORco (*OR-co*receptor); these two units are believed to act in concert to form a ligand-gated nonselective ion channel.[51–53] Whether ORs utilize G-proteins (Fig. 2C) to propagate neuronal signals remains controversial,[51,52,54,55] but it is the OR-mediated depolarization of the cell membrane that triggers action potentials leading to odor recognition and subsequent behaviors.

Ors have been identified in *Ae. aegypti*, *Anopheles gambiae*, *Culex pipiens*, *Anopheles stephensi*, *Anopheles quadriannulatus*, and *G. morsitans*. In mosquitoes, *Ors* mainly express not only in the antenna but also in the maxillary palp, proboscis, and tarsi[11,44,56] in trichoid sensilla as well as in a small number of coeloconic sensilla on the antennae.[57,58] On the maxillary palps, three *Ors* express consistently in the basiconic sensilla of *An. gambiae*[59] and *Ae. aegypti*,[11] the only olfactory structures on this appendage.[60–62]

Mosquito ORs may be broadly tuned or narrowly tuned,[63,64] and some show enhanced sensitivity to select aromatic and heterocyclic compounds

(Fig. 2C) that are components of human odor.[65] However, this distinction may be artificial since deorphanization efforts have relied on a limited array of odorants and a paucity of knowledge of chemical signals used by insect vectors.[66] In addition, non-pheromone receptors, such as those for octenol, skatole, or indole, have sensitivity and selectivity comparable to pheromone receptors.[66–70] Knocking down the expression of ORco prevents *Ae. aegypti* from distinguishing human versus non-human hosts suggesting that a subset of ORs is tuned to host-specific odors.[71]

2.3. Gustatory receptors

GRs appear to be more related to ORco than to other ORs based on protein sequence alignment and gene structure.[72,73] Like ORs, GRs function as heteromers with few exceptions.[74–77] However, GRs differ from ORs where ORco is an obligate coreceptor in that no single conserved coreceptor seems to be involved with functional GR protein complexes.[78] *D. melanogaster* GRs are activated by contact pheromones, sugars and bitter compounds,[30] and are expressed in taste sensilla on the mouthparts and tarsi, an expression pattern also observed for the yellow fever mosquito *Ae. aegypti*.[79] The detection of volatile CO_2 in both *D. melanogaster*[80] and mosquitoes[81] requires two of three highly conserved insect GRs.[82] These CO_2 receptors are positioned in the dendrites of a multidendritic neuron, atypical of *Gr*-expressing neurons, encased in a multiporous basiconic sensillum.[83] In mosquitoes, both the CO_2 and octenol receptor neurons occur in the same basiconic sensilla on the maxillary palp.[84]

Gr genes have been identified in several insect vectors (Fig. 1). The functional CO_2 receptor of *D. melanogaster* consists of two subunits (GR21a and GR63a),[81] and orthologs (*Gr1*, *Gr2*, and *Gr3*) have been identified in mosquitoes[80] and tsetse flies.[17] While GR1 and GR3 are required for CO_2 sensing in mosquitoes, GR2 is not, and its role remains unclear.[82,85] *Gr*-mediated CO_2 sensing requires $G\alpha_q$ in *D. melanogaster*[54] and $G\alpha$-protein genes have characterized in *An. gambiae*,[86] but no functional studies have been carried out on this gene family in insect vectors.

2.4. Ionotropic receptors

IRs are likely more ancient than ORs and GRs as they predate the deuterostome-protostome split[14] and belong to the ionotropic glutamate receptor (iGluR) group found in vertebrates and invertebrates that play roles in synaptic transmission.[87] The mammalian iGluRs likely consist of four or five subunits encoded by one or more genes. In insects, a large portion of this group of proteins is required for normal olfactory function,[88,89] and in at least one instance two IRs are required for a response to a known ligand.[90]

In *D. melanogaster* antennae, *Ir* genes are expressed exclusively in coeloconic sensilla,[88] while no specific cellular expression has been reported for disease vector insects. However, next-generation sequencing of whole appendages shows *Ir* expression in the antennae, maxillary palps, labella, and tarsi of mosquitoes.[44,56,91]

To date, there are no functional studies of mosquito or other insect vector IRs. While some mosquito ORs detect human-specific odorants, IRs may be sensitive to odorants associated with generic vertebrate detection.[71] This proposed function of IRs is consistent with the observation that *ORco* is not expressed in lactic acid-sensitive[92] cells innervating antennal basiconic sensilla of *An. gambiae* and *Ae. aegypti*.[57,58] It is likely that mosquito IRs are sensitive to acids and amines based on functional characterizations of *D. melanogaster* IRs (Fig. 2C).[93]

2.5. Sensory neuron membrane proteins

Insect SNMPs are membrane proteins related to a group of vertebrate scavenger receptors involved in lipid transport, fatty acid uptake, and cell:cell communication.[94,95] SNMPs are thought to play a role in pheromone detection as they are enriched in pheromone-sensitive ORNs in the antennae of moths.[96,97] In *D. melanogaster*, SNMP1 associates with antennal sensilla involved in detection of the pheromone *cis*-vaccenyl acetate (cVA).[98,99] In mosquitoes, *Snmp1* expression is enriched in antennae[100] with virtually no expression in gustatory organs,[44] suggesting an important role for SNMPs in mosquito olfaction. The mechanism of SNMP function in the context of OR-expressing ORNs is unclear, though an increase in spontaneous ORN activity has been observed in SNMP null *D. melanogaster* in addition to cVA insensitivity.[98]

2.6. Activation of olfactory receptor neurons

The molecular events taking place prior to chemosensory receptor (OR, IR, and GR) activation are poorly understood. It is thought that hydrophobic odorants are carried across the aqueous lumen of the sensillum lymph to interact with membrane-bound receptors on the dendrites of ORNs by OBPs and CSPs.[18,101] However, evidence for such a role is indirect and is limited to expression data,[24] OBP/CSP ligand-binding capability *in vitro*, and the requirement of these proteins for normal physiology and behavior. For instance, OBPs and ORs coexpress in the sensilla of *An. gambiae*[102,103] and many of these OBPs express in high concentrations in olfactory appendages of both *An. gambiae* and *Ae. aegypti*.[44,56,104–107] While OBPs and CSPs may be required for the detection of specific odorants,[31,108–110] they often display moderate to high binding affinity to a variety of ligands.[22,43,111–114] Thus, while soluble olfactory proteins are integral to normal olfactory responses, the extent to which they interact with odorants and receptors remains unclear.[32,33]

Since the discovery of ORs,[49] IRs, and GRs too have been shown to play a role in olfaction, demonstrating that insects have evolved independent and parallel solutions to detect odorants. Most evidence indicates that ORs and IRs act as ligand-gated ion channels.[51,52,93] However, there are several reports of G-protein involvement in OR transduction[52,115–117] and at least one pheromone-sensitive OR in the moth *Manduca sexta* acting as a metabotropic receptor.[118]

A parsimonious explanation of receptor activation by an odorant is that ORs and IRs behave like most enzymes, relying on changes in tertiary structure brought on by steric interactions between ligand and ligand-binding regions[119] leading to signal transduction. This "induced-fit" model of receptor activation has not been comprehensively tested in insect ORs or IRs. Nevertheless, it is clear that single amino acid changes that induce structural changes to the receptor can alter odor-induced activation.[53,120,121] Further complicating this proposed mechanism is the possible requirement of consecutive conformational changes, first for an OBP carrier and second for the receptor recognizing the activated OBP, required for the activation of a *D. melanogaster* pheromone OR.[32] An alternative mechanism of OR and IR activation is that receptors exist in a dynamic equilibrium, constantly shifting three-dimensional structure until ligand binding occurs.[122] Despite the uncertainty surrounding receptor activation, receptor molecules are the gateways to the transduction of chemical cues into neural signals mediating attraction and repulsion behaviors, and as such are primary targets for disruption of host-seeking behavior in disease vectors.

While some receptors respond to a relatively narrow range of chemical stimuli, such as relevant pheromones or host odors,[73,123] and other receptors respond to a wide range of odorants, such as plant volatiles,[63,64] these distinctions are difficult to apply generally to receptor proteins. First, response profiles of individual ORs and IRs are often determined through heterologous or *ex vivo* studies, which may or may not reflect the true response profile *in vivo* where nonreceptor molecules like OBPs and SNMPs may affect responses to odorants. Second, the high concentrations of odorants sometimes used to screen receptors are not ecologically relevant and may evoke unnatural responses unrelated to the selected state of the receptor. Testing responses of receptors to a panel of hundreds of odorants at relatively high concentrations may miss key odors that activate receptors at very low concentrations in the nanomolar range. Third, it is difficult to define the odor space in which individual receptors have evolved. Attempts to define a receptor's response range in a laboratory setting are limited by the number and diversity of test odorants.

2.7. Processing of olfactory signals in the brain

The insect olfactory system is capable of detecting and integrating several key odors among a complex blend to form a single perception.[124] These percepts help insects evaluate potential food sources quickly and effectively.[125] Altering an odor cue or creating one that masks or trumps another (through repellent action) can change the perception and behavior of insect vectors. Processing of odor combinations can occur both at the periphery amongst differentially activated ORNs or within the antennal lobe (AL) through interconnected glomeruli.[124] Physiological recordings of electrical activity of olfactory organs can reveal the logic of peripheral coding in disease vectors,[65,126,127] with blends of host cues eliciting different responses than single components. Similarly, downstream processing of odorous blends may be additive or configural (Fig. 3).[128] After detection but before perception, the circuits processing primary odor information reveal target pathways that gate behaviors like host seeking. While repellents do not directly target second- and third-order neurons associated with olfaction, the function of higher brain centers may indicate which ORNs (upstream signals) are most important.

The chemoreceptor repertoire (OR, IR, or GR) of an insect ORN defines its phenotype (Fig. 2).[129] Axons of ORNs of identical phenotype, i.e., those expressing the same receptors, generally coalesce into bundles in the antennal and maxillary nerves to project to spatially segregated glomeruli.[130,131] These glomeruli represent the first level of olfactory integration used by insects to make important behavioral choices (Fig. 3), e.g., whether or not to feed on a particular host organism. The differences between ORN subtypes are evident in the AL, as odorants evoke spatial patterns of neuronal activity[132] allowing researchers a glimpse of the raw discriminatory abilities of many insects.

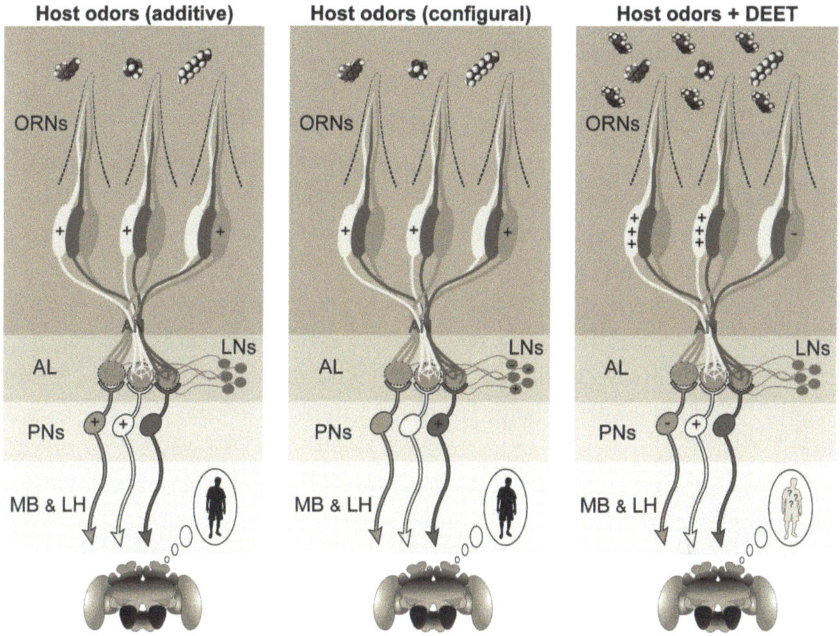

Figure 3 Neural processing of olfactory cues in insect vectors. Odorants stimulate olfactory receptor neurons (ORNs) that have distinct phenotypes, as depicted by yellow, orange, and blue. Axons of ORNs of same phenotype coalesce into a single glomerulus in the antennal lobe (AL) region of the brain. Local interneurons (LNs) form synaptic connections between subsets of AL glomeruli, both stimulating and inhibiting AL responses. Projection neurons (PNs) transmit spatially segregated neural signals of the AL to higher brain centers including the mushroom bodies (MBs) and lateral horn (LH) for further processing. These circuits mediate behavioral responses to odors in insects. The three panels depict olfactory circuit logic possibilities in insects. (+) Indicates neural stimulation, (−) indicates neural inhibition, and (+++) indicates neural hyperstimulation. (Left panel) Additive processing in the AL shows the resulting signals from an odorous blend as the sum of the responses to individual components. (Center panel) Configural processing in the AL shows the resulting signals from an odorous blend as unique and not the sum of the responses to individual components. LNs may mediate AL processing. (Right panel) Repellents like DEET may alter ORN responses to host odors, potentially confusing the insect vector's ability to discriminate a host. These effects change the perception of insects, thus altering host-seeking behavior. (See the color plate.)

The size and arrangement of AL glomeruli vary greatly among insect species,[133] highlighting fundamental differences in odor perception that influence key insect behaviors.[134] For example, honeybees are social insects and have over 150 distinct glomeruli in the AL[135] that respond uniquely to different odors.[136] Even within a single species of honeybee, AL variation among castes reflects differences in behavioral niches.[137] In addition, some social insects have dual second-order pathways that project to the higher brain centers common to all insects, the mushroom bodies (MBs) and lateral horn (LH).[138,139] These dual pathways in honeybees indicate an adaptation related to memory and learning,[140] a function of the MB and LH in insects that remains enigmatic.[141–143] Unique anatomical features of insect brains often reflect unique capabilities, e.g., the macroglomeruli of moths involved in pheromone perception.[144–146]

The arrangement of AL circuits and their response pattern to repellents in insect disease vectors may allow for targeting ORNs for the development of effective alternatives. Both *Ae. aegypti* (50 glomeruli in females, 49 in males[131]) and *An. gambiae* (60 in females, 61 in males[147]) have more AL glomeruli than the hematophagous bug *Rhodnius proxilus* (only 22 glomeruli with less clear demarcation[148]). As these disease vectors use unique combinations of volatiles to locate a bloodmeal, corresponding anatomical features of their olfactory neuropil provide logical targets for behavioral disruption. No AL macroglomerular complexes have been described and only slight AL sexual dimorphism has been reported for mosquitoes.[131,147] It is possible that other features of the mosquito olfactory system such as differences in receptor gene expression account for sexually dimorphic behaviors, e.g., host seeking in blood-feeding females. With recent advances in the genetic modification of insects,[71,149] it may be possible to trace the patterns of activity within the AL elicited by chemical cues, thereby uncovering the logic of host odor/repellent processing in the brain of disease vectors.

The neural output of primary ORNs is carried to the higher brain centers of the LH and MBs in insects by secondary projection neurons (PNs) (Fig. 3). The relationship between number of PNs and AL glomeruli is nearly 1:1,[150] roughly maintaining the spatial divisions of ORN subtypes established in the AL. However, PNs are capable of discriminating primary odor information carried from ORNs. For example, differing concentrations of CO_2 activate the same ORN subtype in *D. melanogaster*, while subsequent PN response patterns mediate avoidance behavior in a concentration-dependent manner.[151] The processing of olfactory signals in higher brain regions remains puzzling, but it is thought to rely on neural oscillations and repeatable firing patterns that are not easily mapped in the MB and LH.[152,153] Nevertheless, it is clear that insects integrate visual, temperature, humidity, and olfactory signals to drive key behaviors such as host seeking.[85,154] Olfactory disruption within higher brain centers would be difficult to target specifically as these regions are not directly exposed to the chemical environment, but the pattern of activity in these regions may highlight the minimum requirements for host-seeking behavior.

2.8. Volatile sensation in GRNs

Insect disease vectors may possess olfactory capabilities outside the antennae and maxillary palps. The morphology of sensilla on the mouthparts and legs of *Ae. aegypti* and other mosquitoes has been described as either uniporous or aporous,[155,156] suggesting olfaction is not a predominant feature. This may not be true in all insects, as the sensation of CO_2 occurs in multiporous sense organs of the mouthparts in Lepidoptera,[157] and taste receptors housed in uniporous sensilla in both the blowfly *Phormia regina*[158] and the moth *M. sexta*[159] respond to high concentrations of volatiles. Furthermore, genes classically associated with olfaction (*Ors* and *Irs*) are expressed in gustatory appendages in mosquitoes[44,57,58] and sensilla on these appendages respond to direct odorant stimulation.[160] Sensitivities to volatiles in gustatory organs should not be overlooked when developing repellents.

2.9. Interaction of repellents with olfactory receptors

Repellents including DEET (*N,N*-diethyl-3-methylbenzamide), the most common active ingredient in insect repellents, elicit avoidance behavior from many insect species[161–165] and can prevent bites from disease spreading insect vectors.[4,80] These compounds may stimulate specific ORNs mediating avoidance behavior or modulate physiological responses to attractive odorants.[92,165–168] While electrophysiological responses of ORNs demonstrate repellents alter incoming neural signals, the dynamics of how these chemicals interact with molecular components in the olfactory lumen is unclear.

Important questions remain about how repellents affect the chemosensory receptors of ORNs. Are repellents activating receptors directly at orthosteric sites by structurally mimicking naturally occurring odorants? Do repellents affect receptor activity through allosteric sites? Do repellents activate neural pathways responding to noxious stimuli or do repellents simply confuse olfactory inputs? Can repellents be predicted through receptor/ligand modeling?

The mode of action of repellents[169] has been an active area of investigation since ORs were shown to mediate olfactory responses to DEET in the *D. melanogaster*[170] and malaria mosquito *An. gambiae*.[171] While DEET was proposed to sequester airborne attractants, thereby blocking odor cues before ORN activation,[165] a recent study did not observe this phenomenon.[121] Repellents picaridin (1-piperidinecarboxylic acid 2-(2-hydroxyethyl)-1-

methylpropylester), IR3535 (ethyl 3-[acetyl(butyl)amino]propanoate), and 2-undecanone also affect the activity of insect ORs expressed heterologously.[172]

Three mosquito ORs (OR2, OR8, and OR10) along with ORco respond to chemical signals[59,63,66–70,121,172] involved in feeding and oviposition.[173–176] These highly conserved mosquito ORs are narrowly tuned to: the human skin attractant octenol (OR8/ORco) or the oviposition attractants indole (OR2/ORco) and skatole (OR10/ORco). DEET affects responses of different OR/ORCO combinations by either direct OR activation or modifying OR responses to other odorants,[170,172] consistent with mosquito ORN responses to DEET that are context dependent.[121,177] Pharmacological investigations show DEET directly activates OR2/ORco while interfering with normal OR8/ORco-mediated responses to octenol.[172] A survey of the effects of structurally diverse repellents on these ORs show that ORs can be directly stimulated by repellents or repellents may disrupt responses of ORs to their cognate agonists.[70] Hence, repellents appear to affect multiple OR assemblages resulting in unnatural stimulation or inhibition of ORNs, sometimes simultaneously. Further expanding the potential mode of action of repellents on ORs is the observation that ORco alone can be activated by the agonist VUAA1[178] (2-(4-ethyl-5-(pyridin-3-yl)-4*H*-1,2,4-triazol-3-ylthio)-*N*-(4-ethylphenyl)acetamide) and VUAA1 analogs[66] or inhibited by trace amines.[179]

Mosquitoes lacking functional OR complexes are undeterred by volatile DEET but show avoidance behavior upon contact.[71] These experiments highlight contact repellency to DEET in mosquitoes that is OR independent as had been demonstrated by the discovery of a GR neuron sensitive to the feeding deterrent quinine and several insect repellents including DEET.[180] Several gene families are candidates for the action of DEET in gustatory appendages of the tarsi and labella.[44,79]

IRs may be required for responses to repellents in insects.[181] In *D. melanogaster*, DEET and structurally similar compounds elicit IR40a-dependent responses from a small population of antennal ORNs and avoidance behavior.[181] It is unclear whether these chemically similar compounds are interacting with the same binding site of IR40a or if other downstream molecular components may be involved. While the *Ir40a* gene is conserved in mosquitoes, lice, and sucking bugs,[181] it is unknown if IR40a-related DEET avoidance is common to these species.

DEET also activates a GRN sensitive to aversive compounds in *D. melanogaster* and this response requires multiple bitter sensing GRs.[182]

The identification of a GRN responsive to DEET and other repellents on the labellum of *Ae. aegypti*[180] correlates well with an earlier report suggesting involvement of the labella in behavioral responses to DEET.[183] GRs are generally not thought to be involved in insect volatile detection except for three highly conserved GRs expressed in the maxillary palps of mosquitoes.[56,59,82] Two of these GRs are required for CO_2 detection in insects,[81,82] an important cue for long distance host-seeking behavior for mosquitoes.[184] A blend of 2,3-butanedione, 1-hexanol, 1-butanal, and 1-pentanal causes potentiation of the CO_2-sensitive ORN expressing these GRs in mosquitoes and this activity disrupts upwind host-seeking behavior in *Ae. aegypti* and *Culex quinquefasciatus*.[185] It is unclear whether the repellent blend acts on CO_2 sensitive GRs, as this ORN also responds to human-emitted odors other than CO_2.[186] The responses to human odors enhance host seeking in *Ae. aegypti* and *An. gambiae* and can be blocked by the repellent ethyl pyruvate.[186] It is possible that the orphan GR paralog related to the functional CO_2 receptors mediates human odor detection in this ORN. Several *Irs* express in the maxillary palp of *Ae. aegypti*,[56] extending the list of receptors through which repellents may exert their action in this ORN.

The key to DEET's efficacy in insects may be its effect on diverse receptor families. It seems unlikely that DEET is capable of interacting with orthosteric sites of receptors whose cognate ligands are chemically dissimilar; rather DEET more likely masks attractive chemical cues through allosteric interactions with receptors for them.[66] Thus, DEET may scramble an evolutionarily selected sensory code by altering odor-gated ion channel dynamics. Single amino acid substitutions in ORs can dramatically change DEET's effect on them,[121] suggesting that the ligand-binding OR subunit may be directly affected. Another common repellent, citronellal, activates TRPA1 (an ion channel involved in many sensory modalities in insects[187]) from *An. gambiae*,[188] thus presenting the possibility that some repellents modulate the activity of other highly conserved ion channels.

Structure-based ligand design can be used to predict effective repellents. *In silico* prediction of repellent activity[189] has led to the identification of moderately effective repellent compounds acting through specific receptors[181] and specific ORNs.[186] The crystal structure of a DEET-bound mosquito OBP has unambiguously demonstrated that a repellent can bind to an olfactory protein.[113] A mosquito OBP bound an attractive compound and a repellent with similar affinity, even simultaneously,[190] suggesting that repellents could disrupt the translocation of host odors across the aqueous lumen of olfactory sensilla (Fig. 2). While it is clear that OBPs influence olfactory driven behaviors of insects including mosquitoes,[31,47,191–193] the usefulness of targeting OBPs for repellent prediction has not been confirmed.

Behavioral insensitivity to DEET based on genetic traits has been observed in insects[194,195] and may be related to the observation that single amino acid changes in an OR can reduce the physiological responses to repellents.[121] Perhaps, blends of repellents may circumvent naturally occurring insensitivity to individual repellents in insect populations and provide opportunities for increased efficacies. It is conceivable that repellent insensitivity traits can be stabilized or enhanced through artificial selection, thus identifying the genes involved.

3. DISCOVERY AND DEVELOPMENT OF NEW REPELLENTS

Discovery of novel repellents involves screening catalogues of chemicals to identify ones that alter insect behavior. Strategies for narrowing the list of possibilities (Fig. 4) range from testing behavioral responses of targeted insects to candidate compounds[196,197] to predicting repellents using complex algorithms.[193,198] Alternate approaches start with a desired phenotype (strong attraction or avoidance) and are followed by identification of the required elements, e.g., a genetic difference or component of an odorous blend.

Initial attempts to identify chemicals for protection from insect bites included chemicals applied to both skin and cloth then tested in arm-in-cage trials requiring years of data collection (Fig. 4A).[196,199] These brute force behavioral screens proved fruitful as DEET was discovered and remains in use for over 60 years. Researchers have also used reverse behavioral screens to identify key components of human odor contributing to decreased attraction.[200]

Now with the molecular characterization of chemosensory proteins, more directed searches are possible by targeting specific receptors. For example, high-throughput screening (HTS) based on specific molecular targets allows for screening of thousands of candidate chemicals for repellent activities prior to verification in behavioral tests (Fig. 4A). HTS using human embryonic kidney cells was used to screen over 100,000 candidate repellents leading to the discovery of the first ORco agonist, VUAA1.[178] While VUAA1 stimulates many ORNs concurrently,[178,201] the usefulness of VUAA1 as a behavioral modifier has only been demonstrated on mosquito larvae.[201]

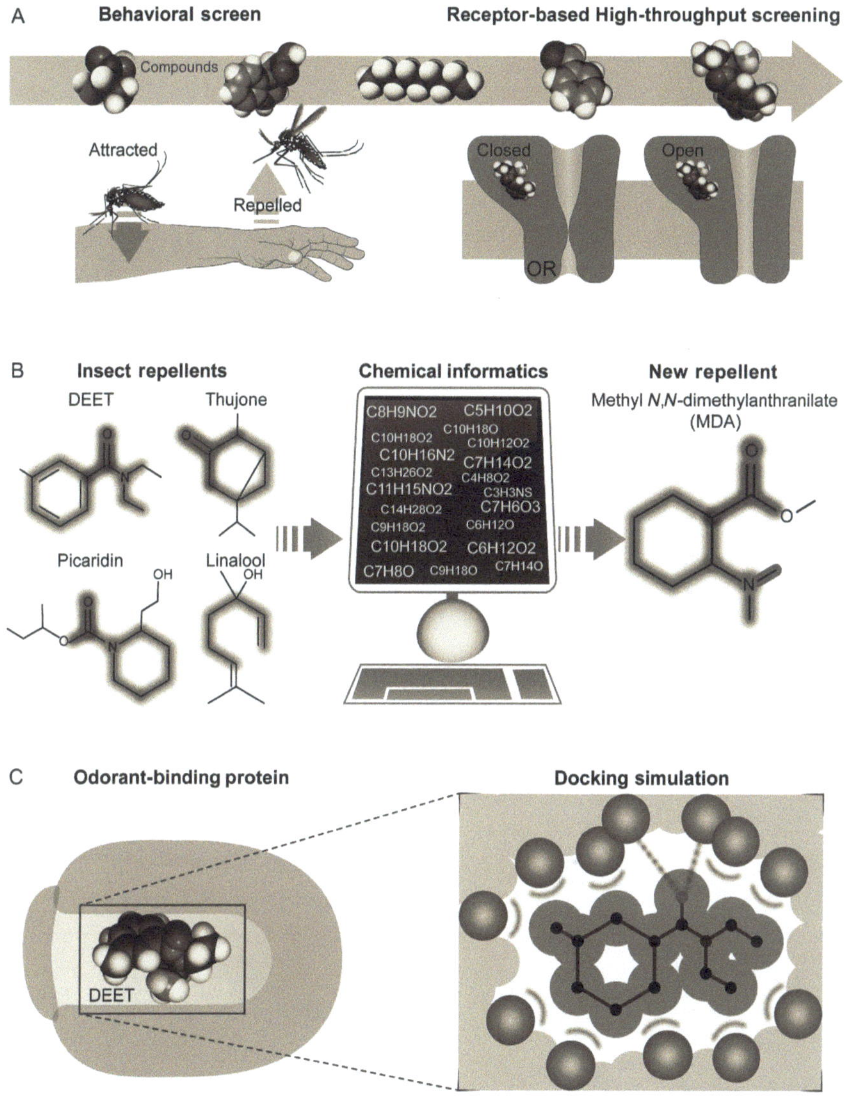

Figure 4 Discovery pipeline of insect repellents. (A) The organism or a specific odorant receptor (OR) might be screened with thousands of chemical compounds to identify behavior- or receptor activity-altering chemicals. (B) Specific chemical descriptors associated with repellency from known insect repellents are established to predict future potential repellents.[181] (C) Amino acid residues interacting with the ligand (here DEET) are mapped using the known crystal structure of an odorant-binding protein. Hydrogen bonds (red dots) and hydrophobic interactions (green arcs) represent these interactions.[113] Mapping of the binding pocket may allow for the rational design of competitive ligand with behavior-modifying properties. (See the color plate.)

A computational approach, based on shared chemical properties of known odorants, has recently been used to predict compounds that elicit responses from specific insect ORs (Fig. 4B).[193] "Chemical informatics" can identify chemicals from a large database (>500,000 structures) with increased likelihoods of repellency in mosquitoes.[181,186] This *in silico* process may itself evolve, incorporating more chemical compounds and more sophisticated algorithms for prediction. Other computational methods target-specific orthosteric sites on receptor molecules for the design of chemicals that initiate specific responses or conformational changes. This rational approach relies on accurate 3D docking models in which either the predicted or crystal structure of a receptor and ligand are analyzed in space and key points of interaction are determined (Fig. 4C).[113,198,202,203]

4. CONCLUSION

Reducing insect olfactory pathways to their molecular components helps identify key interactions between repellents and receptors. DEET remains among the most broadly used repellent for the prevention of insect bites and its effectiveness as a repellent may be due to its effects on a broad spectrum of insect sensory receptors (ORs, IRs, and GRs).[71,131,170,182] However, for DEET and other repellents to have their effects, copious amounts of the chemicals must be applied to either directly to the skin or clothing. Thus, new repellents are needed that are many times more active than DEET and exert their repellent activity at a distance (spatial repellency[204]). Since host seeking in blood-feeding insects is a robust sensory pathway utilizing multiple olfactory and physical cues,[85] novel repellents with modes of action involving an array of sensory targets may be the key for discovery of repellents with greatly improved efficacy.

REFERENCES

1. Gubler DJ. The global emergence/resurgence of arboviral diseases as public health problems. *Arch Med Res*. 2002;33:330–342.
2. Alonso PL, Tanner M. Public health challenges and prospects for malaria control and elimination. *Nat Med*. 2013;19:150–155.
3. Bhatt S, Gething PW, Brady OJ, et al. The global distribution and burden of dengue. *Nature*. 2013;496:504–507.
4. Debboun M, Strickman D. Insect repellents and associated personal protection for a reduction of human disease. *Med Vet Entomol*. 2013;27:1–9.
5. Cruz-López L, Malo EA, Rojas JC, Morgan ED. Chemical ecology of triatomine bugs: vectors of Chagas disease. *Med Vet Entomol*. 2001;15:351–357.

6. Bonilla DL, Durden LA, Eremeeva ME, Dasch GA. The biology and taxonomy of head and body lice: implications for louse-borne disease prevention. *PLoS Pathog.* 2013;9: e1003724.
7. Eisen RJ, Gage KL. Transmission of flea-borne zoonotic agents. *Annu Rev Entomol.* 2012;57:61–82.
8. Moore W, Buc HE. Insect repellent. United States Patent 1,727,305; 1929.
9. Fox AN, Pitts RJ, Robertson HM, Carlson JR, Zwiebel LJ. Candidate odorant receptors from the malaria vector mosquito *Anopheles gambiae* and evidence of down-regulation in response to blood feeding. *Proc Natl Acad Sci USA.* 2001;98:14693–14697.
10. Hill CA, Fox AN, Pitts RJ, et al. G protein-coupled receptors in *Anopheles gambiae. Science.* 2002;298:176–178.
11. Bohbot J, Pitts RJ, Kwon HW, Rützler M, Robertson HM, Zwiebel LJ. Molecular characterization of the *Aedes aegypti* odorant receptor gene family. *Insect Mol Biol.* 2007;16:525–537.
12. Kent LB, Walden KKO, Robertson HM. The Gr family of candidate gustatory and olfactory receptors in the yellow-fever mosquito *Aedes aegypti. Chem Senses.* 2008;33:79–93.
13. Zhou JJ, He XL, Pickett JA, Field LM. Identification of odorant-binding proteins of the yellow fever mosquito *Aedes aegypti*: genome annotation and comparative analyses. *Insect Mol Biol.* 2008;17:147–163.
14. Croset V, Ritz R, Cummins SF, et al. Ancient protostome origin of chemosensory ionotropic glutamate receptors and the evolution of insect taste and olfaction. *PLoS Genet.* 2010;6:e1001064.
15. Liu R, He X, Lehane S, et al. Expression of chemosensory proteins in the tsetse fly *Glossina morsitans morsitans* is related to female host-seeking behaviour. *Insect Mol Biol.* 2012;21:41–48.
16. Members of the International Glossina Genome Initiative. Genome sequence of the tsetse fly (*Glossina morsitans*): vector of African trypanosomiasis. *Science.* 2014;344:380–386.
17. Obiero GF, Mireji PO, Nyanjom SR, Christoffels A, Robertson HM, Masiga DK. Odorant and gustatory receptors in the tsetse fly *Glossina morsitans morsitans. PLoS Negl Trop Dis.* 2014;8:e2663.
18. Vogt RG, Riddiford LM. Pheromone binding and inactivation by moth antennae. *Nature.* 1981;293:161–163.
19. McKenna MP, Hekmat-Scafe DS, Gaines P, Carlson JR. Putative *Drosophila* pheromone-binding proteins expressed in a subregion of the olfactory system. *J Biol Chem.* 1994;269:16340–16347.
20. Pikielny CW, Hasan G, Rouyer F, Rosbash M. Members of a family of *Drosophila* putative odorant-binding proteins are expressed in different subsets of olfactory hairs. *Neuron.* 1994;12:35–49.
21. Li S, Picimbon JF, Ji S, et al. Multiple functions of an odorant-binding protein in the mosquito *Aedes aegypti. Biochem Biophys Res Commun.* 2008;372:464–468.
22. Iovinella I, Bozza F, Caputo B, Della Torre A, Pelosi P. Ligand-binding study of *Anopheles gambiae* chemosensory proteins. *Chem Senses.* 2013;38:409–419.
23. Ozaki M, Morisaki K, Idei W, Ozaki K, Tokunaga F. A putative lipophilic stimulant carrier protein commonly found in taste and olfactory systems: a unique member of the pheromone binding protein family. *Eur J Biochem.* 1995;320:298–308.
24. Galindo K, Smith DP. A large family of divergent odorant-binding proteins expressed in gustatory and olfactory sensilla. *Genetics.* 2001;159:1059–1072.
25. Shanbhag SR, Park SK, Pikielny CW, Steinbrecht RA. Gustatory organs of *Drosophila melanogaster*: fine structure and expression of the putative odorant-binding protein PBPRP2. *Cell Tissue Res.* 2001;304:423–437.

26. Koganezawa M, Shimada I. Novel odorant-binding proteins expressed in the taste tissue of the fly. *Chem Senses*. 2002;27:319–332.
27. Sánchez-Gracia A, Vieira FG, Rozas J. Molecular evolution of the major chemosensory gene families in insects. *Heredity (Edinb)*. 2009;103:208–216.
28. Yasukawa J, Tomioka S, Aigaki T, Matsuo T. Evolution of expression patterns of two odorant-binding protein genes, *Obp57d* and *Obp57e*, in *Drosophila*. *Gene*. 2010;467:25–34.
29. Klein U. Sensillum-lymph proteins from antennal olfactory hairs of the moth *Antherea polyphemus* (Saturniidae). *Insect Biochem*. 1987;17:1193–1204.
30. Jeong YT, Shim J, Oh SR, Yoon HI, Kim CH, Moon SJ. An odorant-binding protein required for the suppression of sweet taste by bitter chemicals. *Neuron*. 2013;79:725–737.
31. Xu P, Atkinson R, Jones DN, Smith DP. *Drosophila* OBP LUSH is required for activity of pheromone-sensitive neurons. *Neuron*. 2005;45:193–200.
32. Laughlin JD, Ha TS, Jones DN, Smith DP. Activation of pheromone-sensitive neurons is mediated by conformational activation of pheromone-binding protein. *Cell*. 2008;133:1255–1265.
33. Gomez-Diaz C, Reina JH, Cambillau C, Benton R. Ligands for pheromone-sensing neurons are not conformationally activated odorant binding proteins. *PLoS Biol*. 2013;11:e1001546.
34. Wojtasek H, Leal WS. Conformational change in the pheromone-binding protein from *Bombyx mori* induced by pH and by interaction with membranes. *J Biol Chem*. 1999;274:30950–30956.
35. Sandler BH, Nikonova L, Leal WS, Clardy J. Sexual attraction in the silkworm moth: structure of the pheromone-binding protein-bombykol complex. *Chem Biol*. 2000;7:143–151.
36. Horst R, Damberger F, Luginbühl P, et al. NMR structure reveals intramolecular regulation mechanism for pheromone binding and release. *Proc Natl Acad Sci USA*. 2001;98:14374–14379.
37. Lee D, Damberger FF, Peng G, et al. NMR structure of the unliganded *Bombyx mori* pheromone binding protein at physiological pH. *FEBS Lett*. 2002;531:314–318.
38. Leal WS, Chen AM, Ishida Y, et al. Kinetics and molecular properties of pheromone binding and release. *Proc Natl Acad Sci USA*. 2005;102:5386–5391.
39. Lautenschlager C, Leal WS, Clardy J. *Bombyx mori* pheromone-binding protein binding nonpheromone ligands: implications for pheromone recognition. *Structure*. 2007;15:1148–1154.
40. Wogulis M, Morgan T, Ishida Y, Leal WS, Wilson DK. The crystal structure of an odorant binding protein from *Anopheles gambiae*: evidence for a common ligand release mechanism. *Biochem Biophys Res Commun*. 2006;339:157–164.
41. Mao Y, Xu X, Xu W, et al. Crystal and solution structures of an odorant-binding protein from the southern house mosquito complexed with an oviposition pheromone. *Proc Natl Acad Sci USA*. 2010;107:19102–19107.
42. Leite NR, Krogh R, Xu W, et al. Structure of an odorant-binding protein from the mosquito *Aedes aegypti* suggests a binding pocket covered by a pH-sensitive "lid". *PLoS One*. 2009;4:e8006.
43. Lagarde A, Spinelli S, Tegoni M, et al. The crystal structure of odorant binding protein 7 from *Anopheles gambiae* exhibits an outstanding adaptability of its binding site. *J Mol Biol*. 2011;414:401–412.
44. Sparks JT, Bohbot JD, Dickens JC. The genetics of chemoreception in the labella and tarsi of *Aedes aegypti*. *Insect Biochem Mol Biol*. 2014;48:8–16.
45. Ozaki M, Takahara T, Kawahara Y, et al. Perception of noxious compounds by contact chemoreceptors of the blowfly, *Phormia regina*: putative role of an odorant-binding protein. *Chem Senses*. 2003;28:349–359.

46. Matsuo T, Sugaya S, Yasukawa J, Aigaki T, Fuyama Y. Odorant-binding proteins OBP57d and OBP57e affect taste perception and host-plant preference in *Drosophila sechellia*. *PLoS Biol.* 2007;5:e118.
47. Harada E, Haba D, Aigaki T, Matsuo T. Behavioral analyses of mutants for two odorant-binding protein genes, *Obp57d* and *Obp57e*, in *Drosophila melanogaster. Genes Genet Syst.* 2008;83:257–264.
48. Swarup S, Morozova TV, Sridhar S, Nokes M, Anholt RRH. Modulation of feeding behavior by odorant-binding proteins in *Drosophila melanogaster. Chem Senses.* 2013;39(2):125–132.
49. Clyne PJ, Warr CG, Freeman MR, Lessing D, Kim J, Carlson JR. A novel family of divergent seven-transmembrane proteins: candidate odorant receptors in *Drosophila. Neuron.* 1999;22:327–338.
50. Benton R, Sachse S, Michnick SW, Vosshall LB. Atypical membrane topology and heteromeric function of *Drosophila* odorant receptors *in vivo*. *PLoS Biol.* 2006;4:e20.
51. Sato K, Pellegrino M, Nakagawa T, Vosshall LB, Touhara K. Insect olfactory receptors are heteromeric ligand-gated ion channels. *Nature.* 2008;452:1002–1006.
52. Wicher D, Schafer R, Bauernfeind R, et al. *Drosophila* odorant receptors are both ligand-gated and cyclic-nucleotide-activated cation channels. *Nature.* 2008;452:1007–1011.
53. Nakagawa T, Pellegrino M, Sato K, Vosshall LB, Touhara K. Amino acid residues contributing to function of the heteromeric insect olfactory receptor complex. *PLoS One.* 2012;7:e32372.
54. Yao CA, Carlson JR. Role of G-proteins in odor-sensing and CO_2-sensing neurons in *Drosophila. J Neurosci.* 2010;30:4562–4572.
55. Ignatious Raja JS, Katanayeva N, Katanaev VL, Galizia CG. Role of $G_{o/i}$ subgroup of G proteins in olfactory signaling of *Drosophila melanogaster. Eur J Neurosci.* 2014;39:1245–1255.
56. Bohbot JD, Sparks JT, Dickens JC. The maxillary palp of *Aedes aegypti* a model of multisensory integration. *Insect Biochem Mol Biol.* 2014;48:29–39.
57. Pitts RJ, Fox AN, Zwiebel LJ. A highly conserved candidate chemoreceptor expressed in both olfactory and gustatory tissues in the malaria vector *Anopheles gambiae. Proc Natl Acad Sci USA.* 2004;101:5048–5063.
58. Melo ACA, Rützler M, Pitts RJ, Zwiebel LJ. Identification of a chemosensory receptor from the yellow fever mosquito, *Aedes aegypti*, that is highly conserved and expressed in olfactory and gustatory organs. *Chem Senses.* 2004;29:403–410.
59. Lu T, Qiu YT, Wang G, et al. Odor coding in the maxillary palp of the malaria vector mosquito *Anopheles gambiae. Curr Biol.* 2007;17:1533–1544.
60. McIver S, Charlton C. Studies on the sense organs on the palps of selected culicine mosquitoes. *Can J Zool.* 1970;48:293–295.
61. McIver SB. Fine structure of pegs on the palps of female culicine mosquitoes. *Can J Zool.* 1972;50:571–576.
62. McIver S, Siemicki R. Palpal sensilla of selected anopheline mosquitoes. *J Parasitol.* 1975;61:535–538.
63. Wang G, Carey AF, Carlson JR, Zwiebel LJ. Molecular basis of odor coding in the malaria vector mosquito *Anopheles gambiae. Proc Natl Acad Sci USA.* 2010;107:4418–4423.
64. Carey AF, Wang G, Su CY, Zwiebel LJ, Carlson JR. Odorant reception in the malaria mosquito *Anopheles gambiae. Nature.* 2010;464:66–71.
65. Cork A, Park KC. Identification of electrophysiologically-active compounds for the malaria mosquito, *Anopheles gambiae*, in human sweat extracts. *Med Vet Entomol.* 1996;10:269–276.
66. Bohbot JD, Dickens JC. Odorant receptor modulation: ternary paradigm for mode of action of insect repellents. *Neuropharmacology.* 2012;62:2086–2095.

67. Bohbot JD, Dickens JC. Characterization of an enantioselective odorant receptor in the yellow fever mosquito *Aedes aegypti*. *PLoS One*. 2009;4:e7032.
68. Pelletier J, Hughes DT, Luetje CW, Leal WS. An odorant receptor from the southern house mosquito *Culex pipiens quinquefasciatus* sensitive to oviposition attractants. *PLoS One*. 2010;5:e10090.
69. Hughes DT, Pelletier J, Luetje CW, Leal WS. Odorant receptor from the southern house mosquito narrowly tuned to the oviposition attractant skatole. *J Chem Ecol*. 2010;36:797–800.
70. Bohbot JD, Jones PL, Wang G, Pitts RJ, Pask GM, Zwiebel LJ. Conservation of indole responsive odorant receptors in mosquitoes reveals an ancient olfactory trait. *Chem Senses*. 2011;36:149–160.
71. DeGennaro M, McBride CS, Seeholzer L, et al. orco mutant mosquitoes lose strong preference for humans and are not repelled by volatile DEET. *Nature*. 2013;498:487–491.
72. Dunipace L, Meister S, McNealy C, Amrein H. Spatially restricted expression of candidate taste receptors in the *Drosophila* gustatory system. *Curr Biol*. 2001;11:822–835.
73. Robertson HM, Warr CG, Carlson JR. Molecular evolution of the insect chemoreceptor gene superfamily in *Drosophila melanogaster*. *Proc Natl Acad Sci USA*. 2003;100:14537–14542.
74. Jiao Y, Moon SJ, Wang X, Ren Q, Montell C. Gr64f is required in combination with other gustatory receptors for sugar detection in *Drosophila*. *Curr Biol*. 2008;18:1797–1801.
75. Moon SJ, Lee Y, Jiao Y, Montell C. A *Drosophila* gustatory receptor essential for aversive taste and inhibiting male-to-male courtship. *Curr Biol*. 2009;19:1623–1627.
76. Lee Y, Moon SJ, Montell C. Multiple gustatory receptors required for the caffeine response in *Drosophila*. *Proc Natl Acad Sci USA*. 2009;106:4495–4500.
77. Weiss LA, Dahanukar A, Kwon JY, Banerjee D, Carlson JR. The molecular and cellular basis of bitter taste in *Drosophila*. *Neuron*. 2011;69:258–272.
78. Isono K, Morita H. Molecular and cellular designs of insect taste receptor system. *Front Cell Neurosci*. 2010;4:1–16.
79. Sparks JT, Vinyard BT, Dickens JC. Gustatory receptor expression in the labella and tarsi of *Aedes aegypti*. *Insect Biochem Mol Biol*. 2013;43:1161–1171.
80. Jones WD, Cayirlioglu P, Kadow IG, Vosshall LB. Two chemosensory receptors together mediate carbon dioxide detection in *Drosophila*. *Nature*. 2007;445:86–90.
81. Erdelyan CNG, Mahood TH, Bader TSY, Whyard S. Functional validation of the carbon dioxide receptor genes in *Aedes aegypti* mosquitoes using RNA interference. *Insect Mol Biol*. 2012;21:119–127.
82. Robertson HM, Kent LB. Evolution of the gene lineage encoding the carbon dioxide receptor in insects. *J Insect Sci*. 2009;9:19.
83. de Bruyne M, Foster K, Carlson JR. Odor coding in the *Drosophila* antenna. *Neuron*. 2001;30:537–552.
84. Grant AJ, Muskavitch MA, O'Connell RJ. Malaria infection does not affect the sensitivity of peripheral receptor neurons in *Anopheles stephensi*. *Parasit Vectors*. 2013;6:134.
85. McMeniman CJ, Corfas RA, Matthews BJ, Ritchie SA, Vosshall LB. Multimodal integration of carbon dioxide and other sensory cues drives mosquito attraction to humans. *Cell*. 2014;156:1060–1071.
86. Rützler M, Lu T, Zwiebel LJ. G_{α} encoding gene family of the malaria vector mosquito *Anopheles gambiae*: expression analysis and immunolocalization of $AG_{\alpha q}$ and $AG_{\alpha o}$ in female antennae. *J Comp Neurol*. 2006;499:533–545.
87. Mayer M. Glutamate receptors at atomic resolution. *Nature*. 2006;440:456–462.
88. Benton R, Vannice KS, Gomez-Diaz C, Vosshall LB. Variant ionotropic glutamate receptors as chemosensory receptors in *Drosophila*. *Cell*. 2009;136:149–162.

89. Rytz R, Croset V, Benton R. Ionotropic receptors (IRs): chemosensory ionotropic glutamate receptors in *Drosophila* and beyond. *Insect Biochem Mol Biol.* 2013;43:888–897.
90. Ai M, Blais S, Park JY, Min S, Neubert TA, Suh GS. Ionotropic glutamate receptors IR64a and IR8a form a functional odorant receptor complex *in vivo* in *Drosophila. J Neurosci.* 2013;33:10741–10749.
91. Pitts RJ, Rinker DC, Jones PL, Rokas A, Zwiebel LJ. Transcriptome profiling of chemosensory appendages in the malaria vector *Anopheles gambiae* reveals tissue- and sex-specific signatures of odor coding. *BMC Genomics.* 2011;12:271.
92. Davis EE, Sokolove PG. Lactic acid-sensitive receptors on the antennae of the mosquito, *Aedes aegypti. J Comp Physiol.* 1976;105:43–54.
93. Abuin L, Bargeton B, Ulbrich MH, Isacoff EY, Kellenberger S, Benton R. Functional architecture of olfactory ionotropic glutamate receptors. *Neuron.* 2011;69:44–60.
94. Vogt RG. Biochemical diversity of odor detection: OBPs, ODEs and SNMPs. In: Blomquist GJ, Vogt RG, eds. *Insect Pheromone Biochemistry and Molecular Biology.* London: Elsevier Academic Press; 2003:391–446.
95. Silverstein RL, Febbraio M. CD36, a scavenger receptor involved in immunity, metabolism, angiogenesis, and behavior. *Sci Signal.* 2009;2:re3.
96. Rogers ME, Sun M, Lerner MR, Vogt RG. Snmp-1, a novel membrane protein of olfactory neurons of the silk moth *Antheraea polyphemus* with homology to the CD36 family of membrane proteins. *J Biol Chem.* 1997;272:14792–14799.
97. Forstner M, Gohl T, Gondesen I, Raming K, Breer H, Krieger J. Differential expression of SNMP-1 and SNMP-2 proteins in pheromone-sensitive hairs of moths. *Chem Senses.* 2008;33:291–299.
98. Benton R, Vannice KS, Vosshall LB. An essential role for a CD36-related receptor in pheromone detection in *Drosophila. Nature.* 2007;450:289–293.
99. Jin X, Ha TS, Smith DP. SNMP is a signaling component required for pheromone sensitivity in *Drosophila. Proc Natl Acad Sci USA.* 2008;105:10996–11001.
100. Vogt RG, Miller NE, Litvack R, et al. The insect SNMP gene family. *Insect Biochem Mol Biol.* 2009;39:448–456.
101. Pelosi P, Zhou JJ, Ban LP, Calvello M. Soluble proteins in insect chemical communication. *Cell Mol Life Sci.* 2006;63:1658–1676.
102. Schymura D, Forstner M, Schultze A, et al. Antennal expression pattern of two olfactory receptors and an odorant binding protein implicated in host odor detection by the malaria vector *Anopheles gambiae. Int J Biol Sci.* 2010;6:614–626.
103. Schultze A, Breer H, Krieger J. The blunt trichoid sensillum of female mosquitoes, *Anopheles gambiae*: odorant binding protein and receptor types. *Int J Biol Sci.* 2014;10:426–437.
104. Justice RW, Dimitratos S, Walter MF, Woods DF, Biessmann H. Sexual dimorphic expression of putative antennal carrier protein genes in the malaria vector *Anopheles gambiae. Insect Mol Biol.* 2003;12:581–594.
105. Li ZX, Pickett JA, Field LM, Zhou JJ. Identification and expression of odorant-binding proteins of the malaria-carrying mosquitoes *Anopheles gambiae* and *Anopheles arabiensis. Arch Insect Biochem Physiol.* 2005;58:175–189.
106. Biessmann H, Nguyen QK, Le D, Walter MF. Microarray-based survey of a subset of putative olfactory genes in the mosquito *Anopheles gambiae. Insect Mol Biol.* 2005;14:575–589.
107. Rinker DC, Zhou X, Pitts RJ, Consortium AGC, Rokas A, Zwiebel LJ. Antennal transcriptome profiles of anopheline mosquitoes reveal human host olfactory specialization in *Anopheles gambiae. BMC Genomics.* 2013;14:749.
108. Pelletier J, Guidolin A, Syed Z, Cornel AJ, Leal WS. Knockdown of a mosquito odorant-binding protein involved in the sensitive detection of oviposition attractants. *J Chem Ecol.* 2010;36:245–248.

109. Biessmann H, Andronopoulou E, Biessmann MR, et al. The *Anopheles gambiae* odorant binding protein 1 (*Agam*OBP1) mediates indole recognition in the antennae of female mosquitoes. *PLoS One.* 2010;5:e9471.
110. Katti S, Lokhande N, González D, Cassill A, Renthal R. Quantitative analysis of pheromone-binding protein specificity. *Insect Mol Biol.* 2013;22:31–40.
111. Xu W, Cornel AJ, Leal WS. Odorant-binding proteins of the malaria mosquito *Anopheles funestus sensu stricto*. *PLoS One.* 2010;5:e15403.
112. Yang G, Winberg G, Ren H, Zhang S. Expression, purification and functional analysis of an odorant binding protein *Aaeg*OBP22 from *Aedes aegypti*. *Protein Expr Purif.* 2011;75:165–171.
113. Tsitsanou KE, Thireou T, Drakou CE, et al. *Anopheles gambiae* odorant binding protein crystal complex with the synthetic repellent DEET: implications for structure-based design of novel mosquito repellents. *Cell Mol Life Sci.* 2012;69:283–297.
114. Gu SH, Wang SY, Zhang XY, et al. Functional characterizations of chemosensory proteins of the alfalfa plant bug *Adelphocoris lineolatus* indicate their involvement in host recognition. *PLoS One.* 2012;7:e42871.
115. Kain P, Chakraborty TS, Sundaram S, Siddiqi O, Rodrigues V, Hasan G. Reduced odor responses from antennal neurons of $G_{q\alpha}$, phospholipase C_{β}, and rdgA mutants in *Drosophila* support a role for a phospholipid intermediate in insect olfactory transduction. *J Neurosci.* 2008;28:4745–4755.
116. Deng Y, Zhang W, Farhat K, Oberland S, Gisselmann G, Neuhaus EM. The stimulatory G_{α}(s) protein is involved in olfactory signal transduction in *Drosophila*. *PLoS One.* 2011;6:e18605.
117. Sargsyan V, Getahun MN, Llanos SL, Olsson SB, Hansson BS, Wicher D. Phosphorylation via PKC regulates the function of the *Drosophila* odorant co-receptor. *Front Cell Neurosci.* 2011;5:5.
118. Nolte A, Funk NW, Mukunda L, et al. *In situ* tip-recordings found no evidence for an Orco-based ionotropic mechanism of pheromone-transduction in *Manduca sexta*. *PLoS One.* 2013;8:e62648.
119. Koshland DE. Enzyme flexibility and enzyme action. *J Cell Comp Physiol.* 1959;54:245–258.
120. Nichols AS, Luetje CW. Transmembrane segment 3 of *Drosophila melanogaster* odorant receptor subunit 85b contributes to ligand-receptor interactions. *J Biol Chem.* 2010;285:11854–11862.
121. Pellegrino M, Steinbach N, Stensmyr MC, Hansson BS, Vosshall LB. A natural polymorphism alters odour and DEET sensitivity in an insect odorant receptor. *Nature.* 2011;478:511–514.
122. Monod J, Wyman J, Changeux JP. On the nature of allosteric transitions: a plausible model. *J Mol Biol.* 1965;12:88–118.
123. Dickens JC. Specialized receptor neurons for pheromones and host plant odors in the boll weevil *Anthonomus grandis* Boh. (Coleoptera: Curculionidae). *Chem Senses.* 1990;15:311–331.
124. Lei H, Vickers N. Central processing of natural odor mixtures in insects. *J Chem Ecol.* 2008;34:915–927.
125. Riffell JA, Lei H, Hildebrand JG. Neural correlates of behavior in the moth *Manduca sexta* in response to complex odors. *Proc Natl Acad Sci USA.* 2009;106:19219–19226.
126. Pappenberger B, Geier M, Boeckh J. Responses of antennal olfactory receptors in the yellow fever mosquito *Aedes aegypti* to human body odours. *Ciba Found Symp.* 1996;200:254–263.
127. Meijerink J, van Loon JJ. Sensitivities of antennal olfactory neurons of the malaria mosquito, *Anopheles gambiae*, to carboxylic acids. *J Insect Physiol.* 1999;45:365–373.

128. Clifford MR, Riffell JA. Mixture and odorant processing in the olfactory systems of insects: a comparative perspective. *J Comp Physiol.* 2013;199:911–928.
129. Vosshall LB, Wong AM, Axel R. An olfactory sensory map in the fly brain. *Cell.* 2000;102:147–159.
130. Couto A, Alenius M, Dickson BJ. Molecular, anatomical, and functional organization of the *Drosophila* olfactory system. *Curr Biol.* 2005;15:1535–1547.
131. Ignell R, Dekker T, Ghaninia M, Hansson BS. Neuronal architecture of the mosquito deutocerebrum. *J Comp Neurol.* 2005;493:207–240.
132. Rodrigues V. Spatial coding of olfactory information in the antennal lobe of *Drosophila melanogaster. Brain Res.* 1998;453:299–307.
133. Strausfeld NJ. *Arthropod Brains: Evolution, Functional Elegance, and Historical Significance.* Cambridge, MA: Harvard University Press; 2012.
134. Rossi Stacconi MV, Hansson BS, Rybak J, Romani R. Comparative neuroanatomy of the antennal lobes of 2 homopteran species. *Chem Senses.* 2014;39:283–294.
135. Arnold G, Masson C, Budharugsa S. Comparative study of the antennal lobes and their afferent pathway in the worker bee and the drone (*Apis mellifera*). *Cell Tissue Res.* 1985;242:593–605.
136. Galizia CG, Sachse S, Rappert A, Menzel R. The glomerular code for odor representation is species specific in the honeybee *Apis mellifera. Nat Neurosci.* 1999;2:473–478.
137. Kropf J, Kelber C, Bieringer K, Rössler W. Olfactory subsystems in the honeybee: sensory supply and sex specificity. *Cell Tissue Res.* 2014;357(3):583–595.
138. Abel R, Rybak J, Menzel R. Structure and response patterns of olfactory interneurons in the honeybee, *Apis mellifera. J Comp Neurol.* 2001;437:363–383.
139. Kirschner S, Kleineidam CJ, Zube C, Rybak J, Grünewald B, Rössler W. Dual olfactory pathway in the honeybee, *Apis mellifera. J Comp Neurol.* 2006;499:933–952.
140. Rössler W, Brill MF. Parallel processing in the honeybee olfactory pathway: structure, function, and evolution. *J Comp Physiol.* 2013;199:981–996.
141. Caron SJ, Ruta V, Abbott LF, Axel R. Random convergence of olfactory inputs in the *Drosophila* mushroom body. *Nature.* 2013;497:113–117.
142. Parnas M, Lin AC, Huetteroth W, Miesenböck G. Odor discrimination in *Drosophila*: from neural population codes to behavior. *Neuron.* 2013;79:932–944.
143. Perisse E, Yin Y, Lin AC, Lin S, Huetteroth W, Waddell S. Different Kenyon cell populations drive learned approach and avoidance in *Drosophila. Neuron.* 2013;79:945–956.
144. Matsumoto SG, Hildebrand JG. Olfactory mechanisms in the moth *Manduca sexta*: response characteristics and morphology of central neurons in the antennal lobes. *Proc R Soc Lond B.* 1981;213:249–277.
145. Rospars JP. Structure and development of the insect antenno-deutocerebral system. *Int J Insect Morphol Embryol.* 1988;17:243–294.
146. Kaissling KE, Hildebrand JG, Tumlinson JH. Pheromone receptor cells in the male moth *Manduca sexta. Arch Insect Biochem Physiol.* 1989;10:273–279.
147. Ghaninia M, Hansson BS, Ignell R. The antennal lobe of the African malaria mosquito, *Anopheles gambiae*—innervation and three-dimensional reconstruction. *Arthropod Struct Dev.* 2007;36:23–39.
148. Barrozo RB, Couton L, Lazzari CR, et al. Antennal pathways in the central nervous system of a blood-sucking bug, *Rhodnius prolixus. Arthropod Struct Dev.* 2009;38:101–110.
149. Bassett AR, Liu JL. CRISPR/Cas9 and genome editing in *Drosophila. J Genet Genomics.* 2014;41:7–19.
150. Sun XJ, Tolbert LP, Hildebrand JG. Synaptic organization of the uniglomerular projection neurons of the antennal lobe of the moth *Manduca sexta*: a laser scanning confocal and electron microscopic study. *J Comp Neurol.* 1997;379:2–20.

151. Lin HH, Chu LA, Fu TF, Dickson BJ, Chiang AS. Parallel neural pathways mediate CO_2 avoidance responses in *Drosophila*. *Science*. 2013;340:1338–1341.
152. Stopfer M. Olfactory processing: massive convergence onto sparse codes. *Curr Biol*. 2007;17:R363–R364.
153. Ito I, Bazhenov M, Ong RC, Raman B, Stopfer M. Frequency transitions in odor-evoked neural oscillations. *Neuron*. 2009;64:692–706.
154. Dekker T, Geier M, Cardé RT. Carbon dioxide instantly sensitizes female yellow fever mosquitoes to human skin odours. *J Exp Biol*. 2005;208:2963–2972.
155. McIver S, Siemicki R. Fine structure of tarsal sensilla of *Aedes aegypti*. *J Morphol*. 1978;155:137–155.
156. Lee RMKW, Craig DA. Fine structure of the sense organs on the labella and labium of the mosquito *Aedes aegypti*. *Open Entomol J*. 2009;3:7–17.
157. Kent KS, Harrow ID, Quartararo P, Hildebrand JG. An accessory olfactory pathway in Lepidoptera: the labial pit organ and its central projections in *Manduca sexta* and certain other sphinx moths and silk moths. *Cell Tissue Res*. 1986;245:237–245.
158. Dethier VG. Sensitivity of the contact chemoreceptors of the blowfly to vapors. *Proc Natl Acad Sci USA*. 1972;69:2189–2192.
159. Städler E, Hanson FE. Olfactory capabilities of the "gustatory" chemoreceptors of the tobacco hornworm larvae. *J Comp Phys*. 1975;104:97–102.
160. Kwon HW, Lu T, Rützler M, Zwiebel LJ. Olfactory responses in a gustatory organ of the malaria vector mosquito *Anopheles gambiae*. *Proc Natl Acad Sci USA*. 2006;103:13526–13531.
161. McIver SB. A model for the mechanism of action of the repellent DEET on *Aedes aegypti* (Diptera: Culicidae). *J Med Entomol*. 1981;18:357–361.
162. Mumcuoglu KY, Galun R, Bach U, Miller J, Magdassi S. Repellency of essential oils and their components to the human body louse, *Pediculus humanus humanus*. *Entomol Exp Appl*. 1996;78:309–314.
163. Kline DL, Bernier UR, Posey KH, Barnard DR. Olfactometric evaluation of spatial repellents for *Aedes aegypti*. *J Med Entomol*. 2003;40:463–467.
164. Debboun M, Frances SP, Strickman D. *Insect Repellents: Principles, Methods, and Uses*. Boca Raton, FL: CRC Press; 2006.
165. Syed Z, Leal WS. Mosquitoes smell and avoid the insect repellent DEET. *Proc Natl Acad Sci USA*. 2008;105:13598–13603.
166. Davis EE. Insect repellents: concepts of their mode of action relative to potential sensory mechanisms in mosquitoes (Diptera: Culicidae). *J Med Entomol*. 1985;22:237–243.
167. Kuthiala A, Gupta RK, Davis EE. Effect of the repellent DEET on the antennal chemoreceptors for oviposition in *Aedes aegypti* (Diptera: Culicidae). *J Med Entomol*. 1992;29:639–643.
168. Boeckh J, Breer H, Geier M, et al. Acylated 1,3-aminopropanols as repellents against bloodsucking arthropods. *Pestic Sci*. 1996;48:359–373.
169. Dickens JC, Bohbot JD. Mini review: mode of action of mosquito repellents. *Pestic Biochem Physiol*. 2013;106:149–155.
170. Ditzen M, Pellegrino M, Vosshall LB. Insect odorant receptors are molecular targets of the insect repellent DEET. *Science*. 2008;319:1838–1842.
171. Xia Y, Wang G, Buscariollo D, Pitts RJ, Wenger H, Zwiebel LJ. The molecular and cellular basis of olfactory-driven behavior in *Anopheles gambiae* larvae. *Proc Natl Acad Sci USA*. 2008;105:6433–6438.
172. Bohbot JD, Dickens JC. Insect repellents: modulators of mosquito odorant receptor activity. *PLoS One*. 2010;5:e12138.
173. Millar JG, Chaney JD, Mulla MS. Identification of oviposition attractants for *Culex quinquefasciatus* from fermented Bermuda grass infusions. *J Am Mosq Control Assoc*. 1992;8:11–17.

174. Meijerink J, Braks MAH, Brack AA, et al. Identification of olfactory stimulants for *Anopheles gambiae* from human sweat samples. *J Chem Ecol.* 2000;26:1367–1382.
175. Takken W, van Loon JJ, Adam W. Inhibition of host-seeking response and olfactory responsiveness in *Anopheles gambiae* following blood feeding. *J Insect Physiol.* 2001;47:303–310.
176. Kline DL, Allan SA, Bernier UR, Welch CH. Evaluation of the enantiomers of 1-octen-3-ol and 1-octyn-3-ol as attractants for mosquitoes associated with a freshwater swamp in Florida, U.S.A. *Med Vet Entomol.* 2007;21:323–331.
177. Takken W, Kline DL. Carbon dioxide and 1-octen-3-ol as mosquito attractants. *J Am Mosq Control Assoc.* 1989;5:311–316.
178. Jones PL, Pask GM, Rinker DC, Zwiebel LJ. Functional agonism of insect odorant receptor ion channels. *Proc Natl Acad Sci USA.* 2011;108:8821–8825.
179. Chen S, Luetje CW. Trace amines inhibit insect odorant receptor function through antagonism of the co-receptor subunit. *F1000Res.* 2014;3:84.
180. Sanford JL, Shields VDC, Dickens JC. Gustatory receptor neuron responds to DEET and other insect repellents in the yellow-fever mosquito, *Aedes aegypti. Naturwissenschaften.* 2013;100:269–273.
181. Kain P, Boyle SM, Tharadra SK, et al. Odour receptors and neurons for DEET and new insect repellents. *Nature.* 2013;502:507–512.
182. Lee Y, Kim SH, Montell C. Avoiding DEET through insect gustatory receptors. *Neuron.* 2010;67:555–561.
183. Bar-Zeev M, Schmidt CH. Action of a repellent as indicated by a radioactive tracer. *J Econ Entomol.* 1959;52:268–269.
184. Gillies MT. The role of carbon dioxide in host-finding by mosquitoes: a review. *Bull Entomol Res.* 1980;70:525–532.
185. Turner SL, Li N, Guda T, et al. Ultra-prolonged activation of CO_2-sensing neurons disorients mosquitoes. *Nature.* 2011;474:87–91.
186. Tauxe GM, MacWilliam D, Boyle SM, Guda T, Ray A. Targeting a dual detector of skin and CO_2 to modify mosquito host seeking. *Cell.* 2013;155:1365–1379.
187. Fowler MA, Montell C. *Drosophila* TRP channels and animal behavior. *Life Sci.* 2013;92:394–403.
188. Kwon Y, Kim SH, Ronderos DS, et al. *Drosophila* TRPA1 channel is required to avoid the naturally occurring insect repellent citronellal. *Curr Biol.* 2010;20:1672–1678.
189. Boyle SM, McInally S, Ray A. Expanding the olfactory code by *in silico* decoding of odor-receptor chemical space. *eLife.* 2013;2:e01120.
190. Murphy EJ, Booth JC, Davrazou F, Port AM, Jones DN. Interactions of *Anopheles gambiae* odorant-binding proteins with a human-derived repellent: implications for the mode of action of n, n-diethyl-3-methylbenzamide (DEET). *J Biol Chem.* 2013;288:4475–4485.
191. Kim MS, Repp A, Smith DP. LUSH odorant-binding protein mediates chemosensory responses to alcohols in *Drosophila melanogaster. Genetics.* 1998;150:711–721.
192. Swarup S, Williams TI, Anholt RR. Functional dissection of odorant binding protein genes in *Drosophila melanogaster. Genes Brain Behav.* 2011;10:648–657.
193. Fan J, Francis F, Liu Y, Chen JL, Cheng DF. An overview of odorant-binding protein functions in insect peripheral olfactory reception. *Genet Mol Res.* 2011;10:3056–3069.
194. Reeder NL, Ganz PJ, Carlson JR, Saunders CW. Isolation of a DEET-insensitive mutant of *Drosophila melanogaster* (Diptera: Drosophilidae). *J Econ Entomol.* 2001;94:1584–1588.
195. Stanczyk NM, Brookfield JF, Ignell R, Logan JG, Field LM. Behavioral insensitivity to DEET in *Aedes aegypti* is a genetically determined trait residing in changes in sensillum function. *Proc Natl Acad Sci USA.* 2010;107:8575–8580.

196. Morton FA, Travis BV, Linduska JP. Results of screening tests with materials evaluated as insecticides, miticides, and repellents at Orlando Laboratory, April, 1942 to April, 1947. In: *US Dept Agr Bur Entomol Plant Quaran*. 1947, E-733.
197. Menger DJ, VAN Loon JJ, Takken W. Assessing the efficacy of candidate mosquito repellents against the background of an attractive source that mimics a human host. *Med Vet Entomol*. 2014;28(4):407–413.
198. Miszta P, Basak SC, Natarajan R, Nowak W. How computational studies of mosquito repellents contribute to the control of vector borne diseases. *Curr Comput Aid Drug Des*. 2013;9:300–307.
199. McCabe ET, Barthel WF, Gertler SI, Hall SA. Insect repellents. III. N, N-diethylamides. *J Org Chem*. 1954;19:493–498.
200. Logan JG, Birkett MA, Clark SJ, et al. Identification of human-derived volatile chemicals that interfere with attraction of *Aedes aegypti* mosquitoes. *J Chem Ecol*. 2008;34:308–322.
201. Taylor RW, Romaine IM, Liu C, et al. Structure-activity relationship of a broad-spectrum insect odorant receptor agonist. *ACS Chem Biol*. 2012;7:1647–1652.
202. Tsitsanou KE, Drakou CE, Thireou T, et al. Crystal and solution studies of the "Plus-C" odorant-binding protein 48 from *Anopheles gambiae*: control of binding specificity through three-dimensional domain swapping. *J Biol Chem*. 2013;288:33427–33438.
203. Gopal JV, Kannabiran K. Studies on interaction of insect repellent compounds with odorant binding receptor proteins by *in silico* molecular docking approach. *Interdiscipl Sci*. 2013;5:280–285.
204. Achee NL, Bangs MJ, Farlow R, et al. Spatial repellents: from discovery and development to evidence-based validation. *Malar J*. 2012;11:164.

CHAPTER FIVE

Pheromone Reception in Moths: From Molecules to Behaviors

Jin Zhang*, William B. Walker†, Guirong Wang*[,1]
*State Key Laboratory for Biology of Plant Diseases and Insect Pests, Institute of Plant Protection, Chinese Academy of Agricultural Sciences, Beijing, PR China
†Chemical Ecology Research Group, Department of Plant Protection Biology, Swedish University of Agricultural Sciences, Alnarp, Sweden
[1]Corresponding author: e-mail address: grwang@ippcaas.cn

Contents

Abstract

Male moths detect and find their mates using species-specific sex pheromones emitted by conspecific females. Olfaction plays a vital role in this behavior. Since the first discovery of an insect sex pheromone from the silkmoth *Bombyx mori*, great efforts have been spent on understanding the sensing of the pheromones *in vivo*. Much progress has been made in elucidating the molecular mechanisms that mediate chemoreception in insects in the past few decades. In this review, we focus on pheromone reception and detection in moths, from the molecular to the behavioral level. We trace the information pathway from the capture of pheromone by male antennae, binding and transportation to olfactory receptor neurons, receptor activation, signal transduction, molecule inactivation, through brain processing and behavioral response. We highlight the impact of recent studies and also provide our insights into pheromone processing.

Progress in Molecular Biology and Translational Science, Volume 130
ISSN 1877-1173
http://dx.doi.org/10.1016/bs.pmbts.2014.11.005

1. INTRODUCTION

Olfaction plays a pivotal role in many aspects of insect behavior, such as foraging, prey avoidance, locating oviposition sites, and mate recognition. Possession of sophisticated olfactory system to detect and interpret odorants in the environment is a prerequisite for survival and reproduction in insects. Understanding of the processes and mechanisms of how chemicals are detected by the antenna, transduced to the brain, and consequently translated into behavior is of great significance. The model of sex pheromone communication in moths, involving the production of sex pheromones by females and the corresponding perception by males, has long been a focus of study in insect olfaction research.[1]

The two sex pheromone components of *Bombyx mori* were identified in the female pheromone glands as bombykol [10,12-(*E*,*Z*)-hexadecadien-1-ol] and bombykal [10,12-(*E*,*Z*)-hexadecadien-1-al] at a prevalence ratio of 10:1.[2,3] Of all moths studied so far, *B. mori* has the simplest known sex pheromone system; bombykol alone is sufficient to trigger pheromone-source seeking behavior in male moths.[4,5] However, other moths usually use a more complicated system wherein blends of several pheromone components are mixed at different concentration ratios.[6,7] Identified moth sex pheromone components have typically been characterized as alcohols, aldehydes, or acetates with long hydrocarbon chains (10–18C) containing 1–3 double bonds with variable positions and geometric isomers.[8,9] The variation of the double bond position and the differentiation of the pheromone component proportions may underlie reproductive isolation barriers between species.[4,10–14] This is exemplified by the two sex pheromone races of the European corn borer moth *Ostrinia nubilalis*, E and Z, which produce opposite ratios of isomers of the major female sex pheromone components, E11- and Z11–14:acetate.[15,16] Lassance *et al.* found that allelic variation in a fatty-acyl reductase gene that is essential for pheromone biosynthesis accounts for the female pheromone variation.[17]

In this review, we focus on pheromone reception and detection in moths, from the molecular to the behavioral level. We trace the information pathway from the capture of pheromone by male antennae, binding and transportation to olfactory receptor neurons (ORNs), receptor activation, signal transduction, molecule inactivation, and behavioral response.

2. STRUCTURE OF ANTENNAE

Moths detect odorants primarily in their antennae. On the surface of the antennae, there are thousands of olfactory sensilla and at least five structural types have been categorized based on the morphology of the sensilla: long trichodea, short trichodea, basiconica, coeloconica, and chaetica.[18,19] Previous studies on lepidopterans have demonstrated that in males the long sensilla trichodea are mainly involved in pheromone detection,[3,4,20–24] the short sensilla trichodea are tuned to both the pheromone components and other chemicals,[25] and the sensilla basiconica mainly respond to plant volatiles.[26,27] Electroantennography of *Heliothis virescens* females showed responses to many components of the female sex pheromone.[28] In female *B. mori* moths, neurons in the long sensilla trichodea were shown to respond to the plant volatiles benzoic acid and linalool.[29] Several studies have shown the responsiveness of long sensilla trichodea neurons in female antennae to conspecific sex pheromone.[24,30–32]

In this review, we mainly discuss the pheromone-sensing long trichodea which are usually the longest and most abundant type of sensillum.[33,34]

3. ANTENNAL ORNs

Previous studies have shown that a majority of the ORNs located on male antennae are specifically tuned to only one or a few pheromone components of the conspecific females.[25,35] Various single sensillum recording (SSR) results have indicated that each trichodea sensillum is innervated by the dendrites of two to four ORNs,[24,28,36,37] which also project their axons into the antennal lobe (AL). Three types of non-neuronal cells: trichogen, thecogen, and tormogen cells surround the body of the ORNs. Ljungberg *et al.* identified two physiological types of sensilla trichodea in antennae of male *Spodoptera littoralis*. Sensillum type (I) housed two ORNs (one with large-amplitude spikes, the other with small-amplitude spikes), with the large-spiking one tuned to the secondary pheromone component, Z9E12–14:OAc, and the small-spiking one tuned to a behavioral antagonist, Z9–14:OH. Sensillum type (II) contained one ORN tuned to the major pheromone component, Z9Ell-14:OAc, and a second neuron with no response to any of the pheromone components tested.[24,25] Cossé *et al.* found that three types of sensilla trichodea were located at the lateral edge of the antennomeres of *Helicoverpa zea* male antennae. Type A sensilla (71%)

contained an ORN tuned to the principal sex pheromone component Z11–16:Ald. Type B sensilla (10%) contained an ORN that responded only to Z9–14:Ald. Type C sensilla (19%) contained two ORNs (large-spiking one and small-spiking one), with the large-spiking one tuned to Z9–16: Ald and Z9–14:Ald, while the small-spiking one responded to Z11–16: OAc, Z11–16:OH, and Z9–14:Ald.[36,38,39] Similar ORN distributions were also reported in *Heliothis subflexa* and *H. virescens*.[39]

4. MOLECULAR COMPONENTS OF CHEMICAL RECEPTION

Chemical volatiles are first captured by the antennae of the moths and then absorbed through the pores on the surface of the sensilla.[4,40] Considering that volatile odorants are highly hydrophobic, as well as the existence of odorant-degrading enzymes (ODEs) in the sensillum lymph, odorants do not pass easily through the sensillum lymph surrounding the dendritic membrane of ORNs. However, when bound to odorant-binding proteins (OBPs) in the sensillum lymph, they become solubilized and presumably sheltered from ODE breakdown. When the odorant/ionotropic receptors (ORs/IRs) expressed on the dendritic membrane of ORNs are activated by the odorants alone or by the odorant–OBPs complex, the chemical signals are translated to electrical signals that are transduced to the AL. After activating ORs/IRs, the molecules are inactivated either by enzymatic degradation or via other unidentified inactivation mechanisms.[41–43] The proteins involved in pheromone reception will be discussed in greater detail in the following sections.

5. PHEROMONE RECEPTORS

Pheromones delivered to the dendritic membrane of ORNs are recognized by pheromone receptors (PRs), which are a subclass of insect ORs. Insect ORs were first identified in *Drosophila melanogaster* in 1999.[44,45] The first insect PR was characterized from *B. mori* using heterologous expression in *Xenopus* oocytes.[46] Since then, much progress has been made in identifying new PRs in moths, with characterizations of receptors in more than 15 species.[46–61]

Recently, studies on PRs have mainly involved functional characterization of PRs from different moth species by heterologous expression in HEK293 cells,[49,51] *Xenopus* oocytes,[46] or *Drosophila* antennal transgenic expression systems.[62] The first PRs in insects were identified in *B. mori*, BmorOR1 and BmorOR3. Oocytes expressing BmorOR1/BmorOrco and BmorOR3/BmorOrco responded specifically to the sex pheromone components bombykol and bombykal, respectively.[46,47] However, when expressed in modified HEK293 cells, BmorOR1 responded to both bombykol and bombykal, although with different sensitivity and selectivity, and BmorOR3 responded to bombykal in a dose-dependent manner but not to bombykol.[63]

The European corn borer, *O. nubilalis* (Hubner), has provided a model system to study the evolution of sex pheromones among closely related races (Z and E races) and species. Wanner *et al.* characterized five PRs in *O. nubilalis* (Z), via expression in *Xenopus* oocytes. In this functional study, OnubOR6 was narrowly tuned to the main component Z11–14:OAc, whereas the others, OnubOR1, OnubOR3, and OnubOR5, responded to all four pheromone components tested as well as a behavioral antagonist. OnubOR4 only had a small response to Z9–14:OAc.[55] Seven candidate PRs were identified from *O. scapulalis* (OscaOR1, OscaOR3–8). When coexpressed with OscaOrco in *Xenopus* oocytes, OscaOR1 specifically responded to E11–14:OH, a sex pheromone component of the congener *Ostrinia latipennis*.[52] In a separate study, OscaOR3 was broadly tuned to Z11–14:OAc and E11–14:OAc as well as pheromone components of both congeners. OscaOR4 responded to the main pheromone component E11–14:OAc with a relatively high specificity, which was different from OnubOR4.[53]

The tobacco budworm, *H. virescens*, is another model species that has been extensively studied with respect to pheromone reception and response. In this species, at least seven compounds are released from the female sex pheromone glands, and Z11–16:Ald together with Z9–14:Ald elicit oriented male flight behavior.[64,65] Krieger *et al.* first identified six *H. virescens* PRs expressed in males.[48] Functional characterizations of the PRs were conducted using heterologous expression in *Xenopus* oocytes. HvirOR6 was found to be highly tuned to Z9–14:Ald, while HvirOR13, HvirOR14, and HvirOR16 showed specificity for Z11–16:Ald, Z11–16:OAc, and Z11–16:OH, respectively. HvirOR15 did not respond to any of the pheromone compounds tested.[54] There are some inconsistencies when comparing the results in *Xenopus* oocytes to SSR on long sensilla trichodea on the moth antennae. Type (A) sensilla of *H. virescens* responded specifically to Z11–16:Ald and exhibited no activity to any of the other pheromone-related compounds. Type (B) sensilla responded only to the secondary pheromone component Z9–14:Ald and was responsive to a low threshold of 2 pg. In the *Xenopus* oocytes, HvirOR13, which is expressed in sensillum type (A), responded strongly to Z11–16:Ald, but only slightly to Z9–14:Ald. Besides Z9–14:Ald, HvirOR6 expressed in sensillum type (B) was widely tuned to other pheromones tested. These inconsistencies in tuning profile, sensitivity, and recovery time between the two systems may be due to a lack of essential perireceptor components, such as pheromone-binding proteins (PBPs), general odorant-binding proteins (GOBPs), and sensory neuron membrane proteins (SNMPs), in the *in vitro* system. Each of these perireceptor components is discussed in following sections.

D. melanogaster antennae have proven to be a viable system for transgenic expression and functional characterization of moth PRs. For this strategy, three different approaches have been developed, the first two of which rely upon SSR assays. First, expression of *B. mori* PR, BmorOR1, in the empty neuron system[66] conferred responsiveness of the empty neuron to bombykol[62]; this response was enhanced when BmorPBP was also expressed in the empty neuron sensilla environment. This study was followed up with another showing that expression of BmorOR1 in a sensilla trichodea type (I) ORN resulted in a more highly sensitive response to bombykol than was previously seen in the empty neuron system; presumably this was due to a higher degree of similarity in the perireceptor environment in the type (I) sensilla, as these sensilla contain an ORN/OR responsive to the *D. melanogaster* pheromone *cis*-vaccenyl acetate.[67] Finally, the *S. littoralis* putative PR, SlitOR6, was shown to be responsive to a *S. littoralis* minor pheromone component (*Z,E*)-9,12-tetradecadienyl acetate, when expressed in a majority of *D. melanogaster* antennal olfactory neurons and assayed with electroantennography. For this system, SlitOR6 expression was driven by the DmelORCo promoter, and results were confirmed with SSR assays performed on type (I) sensilla expressing SlitOR6, as referenced above.[56]

The identification and initial functional characterization of the *B. mori* bombykal and bombykol receptors led to the classification of the clade of ORs containing these receptors as the "Moth Pheromone Receptor" clade.[47,48,54,68] This insight spurred significant research on functional characterizations of putative PRs, as described above. However, a couple of studies have shown that some receptors within this clade do not respond to pheromone compounds.[54,58] EposOR1, from the light brown apple moth, *Epiphyas postvitanna*, belonging to the PR clade, was found to be responsive to several plant volatiles when expressed in *Sf*9 cells. These plant compounds were reported to elicit an EAG response from the antennae of *E. postvittana*.[69] EposOR1 was found to be very sensitive to a known moth oviposition deterrent, methyl salicylate and insect repellent, geraniol and citral.[70–72] Recently, the codling moth (*Cydia pomonella*) OR, CpomOR3, was found to be responsive to a compound derived from fruits (including apples and pears) known as pear ester (Ethyl decadienoate),[73] which is attractive to both sexes of adults and larvae of the codling moth.[74,75]

6. PHEROMONE-BINDING PROTEINS

PBPs make up a subfamily of OBPs in insects, which were first identified in male silkmoth, *Antheraea polyphemus*.[76] Since then, PBPs from more than 16 lepidopteran species have been identified.[77–81] PBPs are small (15–20 kDa), water-soluble proteins that are synthesized by two olfactory accessory cells: trichogen and tormogen cells and are secreted abundantly into the sensillum lymph of trichoid sensilla.[82–84] Previous functional studies of PBPs indicate that they may: (a) act as solubilizers and carriers of the hydrophobic pheromones in the aqueous sensillum lymph[63,76,85–90]; (b) concentrate odorants in the sensillum lymph[91]; (c) protect the pheromone from enzymatic degradation; (d) serve as cofactors in the activation of PRs; and (e) be involved in the postulated odorant molecule deactivation.[88,92]

BmorPBP1 from *B. mori* is one of the best-characterized biochemical and structural examples of a PBP. It is expressed in the lymph of the pheromone-sensitive long sensilla trichodea and specifically binds to bombykol rather than bombykal.[84,93] BmorPBP1 binds bombykol with high affinity at the sensillar lymph pH (=7) but shows no affinity when the pH becomes more acidic (=4.5).[94] It has been predicted that when the PBP/pheromone complex is transported to the vicinity of ORNs where the pH is acidic, the pheromones will be released to the dendritic membrane of the ORNs, where they can activate their cognate receptors.[94] The pH-dependent conformational changes in PBPs that result in the release of pheromones have been confirmed structurally and biochemically in several species.[42,94–100] This pattern is consistent with *in vitro* studies showing that pheromones can activate the ORNs directly without the participation of PBPs.[47,101]

However, a strong conflict with this observation is seen in *D. melanogaster*, with the LUSH OBP.[102] In LUSH mutants, ORNs in the type (I) trichoid sensilla are not responsive to the aggregation pheromone 11-*cis*-vaccenyl acetate (cVA). However, when LUSH expression is rescued or recombinant LUSH protein is added into type (I) trichoid sensilla, the response to cVA is restored.[102] Laughlin *et al.* showed that LUSH undergoes a pheromone-specific conformational change that triggers the firing of pheromone-sensitive neurons.[89] These studies concluded that LUSH is a prerequisite for the detection of cVA. However, recent research was inconsistent with these results, showing that there is no conformational property of cVA/LUSH that can explain its proposed unique activated state, and high concentrations of cVA can induce neuronal activity without the presence of LUSH.[103]

Some attempts have been made to investigate the interplay of PBPs and PRs in the past decades. When dissolved in DMSO, both bombykol and bombykal can activate BmorOR1-expressing HEK293 cells. However, when dissolved in BmorPBPs without DMSO, the cells respond to bombykol but not to bombykal.[63] This study indicates that PBPs can solubilize, bind, and interact specifically and differentially with various pheromone compounds.[104–106] Grosse-Wilde *et al.* reported that the response of HvirOR13 (*H. virescens*) to DSMO-solubilized Z11–16:Ald was significantly increased by substituting DMSO with HvirPBP2.[49] These data indicate a specific interplay between HvirOR13, HvirPBP2, and Z11–16:Ald. The enhancement of PBPs in pheromone sensing was also confirmed in *A. polyphemus* using calcium imaging.[51] Our study on interplay between PRs and PBPs from *Plutella xylostella* showed that PxylPBPs could increase the sensitivity of the PR-expressing oocytes to the ligand pheromone component while decreasing the sensitivity to pheromone analogs.[59] However, Xu *et al.* found that when bombykol and bombykal are solubilized with BmorPBP1, there was no enhancement in selectivity; by contrast, both ligands were trapped by BmorPBP1 leading to dramatically reduced responses.[107]

How the PBPs function in pheromone sensing, whether the PBPs release the pheromones to activate the ORNs alone or instead as a complex, is still uncertain. Recently, such functional studies have concentrated mainly on binding experiments and the PR/PBP interplay explorations have been performed *in vitro*. In the coming years, greater focus should be given to *in vivo* functional studies of PBPs and interactions between PBPs and PRs.

7. GENERAL ODORANT-BINDING PROTEINS

GOBPs are another subfamily of OBPs in insects and were named based on their expression in the antennae of both sexes. GOBPs can be subdivided into two groups: GOBP1 and GOBP2 and are more conserved than PBPs across lepidopteran species.[79,108,109] Immunocytochemical localization experiments in *A. polyphemus* showed the anti-GOBP serum labeled almost all the sensilla basiconica in male and female antennae.[108] In another study of *Helicoverpa armigera*, Wang *et al.* found that HarmGOBP2 is mainly expressed in sensilla basiconica in the male antennae, while in the female it is equally expressed in sensilla basiconica and sensilla trichodea.[110] Because of the predominant presence in sensilla basiconica, they were thought to detect general odorants[83,111]; however, the GOBPs may have other functions. In the functional characterization of GOBP2 from *Mamestra brassicae*, MbraGOBP2 bound specifically to the behavioral antagonist Z11–16: OH, with no affinity for the pheromone components Z11–16:Ac, 16:Ac, and Z11–18:Ac. *In situ* hybridization showed MbraGOBP2 was associated with only sensilla trichodea in male antennae.[112] It was suggested that MbraGOBP2 may ensure the specificity of pheromone communication and avoid inbreeding of closely related species. Functional analysis of GOBP2 in *Chilo suppressalis* showed that CsupGOBP2 had significant affinity to the main pheromone component Z11–16:Ald, and to laurinaldehyde and benzaldehyde, two general plant volatile aldehydes.[113] Recently, BmorGOBP2, AtraGOBP2 (*Amyelois transitella*), and LstiGOBP2 (*Loxostege sticticalis*) also have been found to bind to the sex pheromone of the species in which they occur.[114–116] GOBP2 can bind to the sex pheromone in some cases, but the functions of GOBP1 remain largely unknown. Previous studies on GOBPs provide good basic information; however, determining the exact role of GOBPs in pheromone detection requires future *in vivo* studies.

8. SENSORY NEURON MEMBRANE PROTEINS

SNMPs are insect membrane proteins that are associated with pheromone-sensitive neurons in Lepidoptera and Diptera.[117–120] SNMPs belong to the human fatty acid transporter (FAT), CD36 gene family, which is characterized by two transmembrane domains and mainly involved in the recognition of fatty acids, cholesterol, and proteinaceous compounds in cells.[117,121–124] The insect SNMP family consists of two subfamilies, SNMP1 and SNMP2, which were first identified from *A. polyphemus*[117] and *Manduca sexta,*[125] respectively. Since then, much progress has been achieved in the identification of SNMP1 and SNMP2 in different insect orders.[103,119,126–133] ApolSNMP1 mRNA expression increased significantly 1–2 days before adult emergence, coinciding with the functional maturation of the olfactory system. The abundant expression of SNMP1 in pheromone-specific olfactory neurons suggests it may be involved in pheromone detection.[117,126,134] Forstner *et al.* found that in male antennae of *H. virescens*, HvirSNMP1 and HvirOR13 were coexpressed in the same cells, while in contrast, HvirSNMP2 was expressed in the supporting cells. Similar expression profiles of ApolSNMP1 and ApolSNMP2 were found in *A. polyphemus.*[120] DmelSNMP1 has been identified to be a prerequisite for the chemosensory detection of the fatty acid pheromone cVA,[119,135] perhaps similar to the reported function of CD36 proteins in mammals.[136,137] This was the first demonstration of SNMP functions *in vivo*. Additionally, it has been reported that SNMPs display wide expression patterns in different tissues,[129–133] suggesting that they may be involved in some functions beyond olfactory detection.

9. ANTENNAL LOBE

Pheromone signals transduced into electrical signals by ORNs are transmitted to the AL, which is the primary olfactory center of the insect brain, through the axons of ORNs. All synaptic contacts between ORNs and interneurons take place in AL glomeruli.[138,139] In male moth brains, the glomeruli can be divided into two parts: the macroglomerular complex (MGC) located dorsally and the ordinary glomeruli (OG) located ventrally. The MGC is the first center for pheromone synaptic processing in the male AL. In *B. mori*, there are three compartments in the MGC named the cumulus, toroid, and horseshoe.[43] In species of *Helicoverpa* as well as *S. littoralis*, there are three MGC compartments, and in *Agrotis segetum* and *Heliothis* spp. there are four.[24,36,140–143] Three types of AL neurons have branches within the glomeruli: local interneurons that connect to the glomeruli, projection neurons that receive input from the glomeruli and send processed signals to the brain, and centrifugal neurons of unknown significance.[144] In *B. mori*, BmorOR1- and BmorOR3-expressing neurons, responding to bombykol and bombykal, project to the toroid and cumulus, respectively.[145] In *M. sexta*, the ORNs responding to the principal pheromone component bombykal project to the toroid, whereas the ORNs responding preferentially to stimulation with a secondary pheromone component project to the cumulus.[146,147] However, studies in several heliothine species have demonstrated that the cumulus is a common site for processing information derived from the major pheromone component.[36,37,140,142,148–151]

OGs consist of a group of (~60) small-sized glomeruli situated ventrally in the moth AL.[152–155] Each OG receives inputs from the ORNs located beneath olfactory sensilla. The number of glomeruli within the ALs correlates with the number of OSN functional types in the antennae and in the maxillary palps.[156–159] As a result of this, the number of antennally expressed receptors can be estimated based on the total number of glomeruli.

10. BEHAVIOR

When receiving and detecting pheromone signals from conspecific females, male moths exhibit a zigzag upwind flight pattern to the source female. For *B. mori*, the principle pheromone bombykol, alone, is enough to elicit the male mating behavior, while the second pheromone, bombykal, suppresses the behavioral response to bombykol, but the exact function of bombykal remains unknown.[3,43,160] Unlike *B. mori*, females of many other moth species use blends of pheromones, with species-specific ratios of several components, to attract conspecific males. For *H. armigera*, more than seven pheromone components have been identified from the pheromone glands of females.[14,161,162] When the pheromone components Z11–16: Ald and Z9–16:Ald are mixed at a ratio of 99:1 as compared to 90:10, attractiveness to the males is significantly increased.[161,163] Z9–14:Ald was shown to strengthen the attraction of males at low concentrations, while suppressing attraction at higher concentrations, when mixed with the pheromone components.[162]

Due to the high specificity as well as the hardwired predictability of the male moth response to female-produced sex pheromone, mass trapping applications have been successfully developed and implemented to lure male moths for purposes of pest monitoring and sustainable integrated pest management.[164] Continued research on all aspects of moth sex PR biology, with specific emphasis on molecular mechanisms, will facilitate a better understanding of the behavior of moths and a more informed approach to management of agriculturally important pest moths, which inflict significant damage every year on agricultural crops throughout the world.

ACKNOWLEDGMENTS

This work was supported by National Natural Science Foundation of China (31230062 and 31071752) and the China National "973" Basic Research Program (2012CB114104) to G. W.

REFERENCES

1. Karlson P, Butenandt A. Pheromones (ectohormones) in insects. *Annu Rev Entomol.* 1959;4:39–58.
2. Butenandt A, Beckmann R, Stamm D, Hecker E. Über den Sexuallockstoff den Seidenspinners *Bombyx mori.* Reindarstellung und Konstitution. *Z Naturforsch C.* 1959;14b:283–284.
3. Kaissling KE, Kasang G, Bestmann HJ, Stransky W, Vostrowsky O. A new pheromone of the silkworm moth *Bombyx mori. Naturwissenschaften.* 1978;65:382–384.
4. Kaissling KE, Colbow K. *R. H. Wright Lectures on Insect Olfaction.* Burnaby, BC: Simon Fraser University; 1987.
5. Schneider D. 100 years of pheromone research. *Naturwissenschaften.* 1992;79:241–250.
6. Byers JA. Pheromone component patterns of moth evolution revealed by computer analysis of the Pherolist. *J Anim Ecol.* 2006;75:399–407.
7. Smadja C, Butlin RK. On the scent of speciation: the chemosensory system and its role in premating isolation. *Heredity (Edinb).* 2009;102:77–97.
8. Roelofs WL, Rooney AP. Molecular genetics and evolution of pheromone biosynthesis in Lepidoptera. *Proc Natl Acad Sci USA.* 2003;100:9179–9184.
9. Ando T, Inomata S, Yamamoto M. Lepidopteran sex pheromones. *Top Curr Chem.* 2004;239:51–96.
10. Tamaki Y, Noguchi H, Yushima T. Sex pheromone of *Spodoptera litura* (F.) (Lepidoptera: Noctuidae): isolation, identification, and synthesis. *Appl Entomol Zool.* 1973;8:200–203.
11. Tamaki Y, Yushima T. Sex pheromone of the cotton leafworm, *Spodoptera littoralis. J Insect Physiol.* 1974;20:1005–1014.
12. Dong SL, Du JW. Chemical identification and field tests of sex pheromone of beet armyworm *Spodoptera exigua. J Plant Prot.* 2002;29:19–24.
13. Cork A, Boo KS, Dunkelblum E, et al. Female sex pheromone of oriental tobacco budworm, *Helicoverpa assulta* (Guenee) (Lepidoptera: Noctuidae): identification and field testing. *J Chem Ecol.* 1992;18:403–418.
14. Wu D, Yan Y, Cui J. Sex pheromone components of *Helicoverpa armigera*: chemical analysis and field tests*. *Insect Sci.* 1997;4:350–356.
15. Klun JA, Chapman OL, Mattes KC, et al. Insect sex pheromones: minor amount of opposite geometrical isomer critical to attraction. *Science.* 1973;181:661–663.
16. Gomez-Diaz C, Benton R. The joy of sex pheromones. *EMBO Rep.* 2013;14:874–883.
17. Lassance JM, Groot AT, Lienard MA, et al. Allelic variation in a fatty-acyl reductase gene causes divergence in moth sex pheromones. *Nature.* 2010;466:486–489.
18. Faucheux MJ. Morphology and distribution of antennal sensilla in the female and male clothes moth, *Tineola bisselliella* Humm. (Lepidoptera: Tineidae). *Can J Zool.* 1985;63: 355–362.
19. Cornford ME, Rowley WA, Klun JA. Scanning electron microscopy of antennal sensilla of the European corn borer, *Ostrinia nubilalis. Ann Entomol Soc Am.* 1973;66:1079–1088.
20. Schneider D. Insect olfaction: deciphering system for chemical messages. *Science.* 1969;163:1031–1037.

21. Steinbrecht RA. The fine structure of olfactory sensilla in the silk moth (Insecta, Lepidoptera). Receptor processes and stimulus conduction apparatus. *Z Zellforsch Mikrosk Anat.* 1973;139:533–565.
22. Sanes JR, Hildebrand JG. Structure and development of antennae in a moth, *Manduca sexta. Dev Biol.* 1976;51:280–299.
23. Lee JK, Strausfeld NJ. Structure, distribution and number of surface sensilla and their receptor cells on the olfactory appendage of the male moth *Manduca sexta. J Neurocytol.* 1990;19:519–538.
24. Ochieng SA, Anderson P, Hansson BS. Antennal lobe projection patterns of olfactory receptor neurons involved in sex pheromone detection in *Spodoptera littoralis* (Lepidoptera: Noctuidae). *Tissue Cell.* 1995;27:221–232.
25. Ljungberg H, Anderson P, Hansson BS. Physiology and morphology of pheromone-specific sensilla on the antennae of male and female *Spodoptera littoralis* (Lepidoptera: Noctuidae). *J Insect Physiol.* 1993;39:253–260.
26. Lopes O, Barata EN, Mustaparta H, Araújo J. Fine structure of antennal sensilla basiconica and their detection of plant volatiles in the eucalyptus woodborer, *Phoracantha semipunctata* Fabricius (Coleoptera: Cerambycidae). *Arthropod Struct Dev.* 2002;31:1–13.
27. Anderson P, Hansson BS, Löfqvist J. Plant-odour-specific receptor neurones on the antennae of female and male *Spodoptera littoralis. Physiol Entomol.* 1995;20:189–198.
28. Almaas T, Mustaparta H. *Heliothis virescens*: response characteristics of receptor neurons in sensilla trichodea type 1 and type 2. *J Chem Ecol.* 1991;17:953–972.
29. Heinbockel T, Kaissling K-E. Variability of olfactory receptor neuron responses of female silkmoths (*Bombyx mori* L.) to benzoic acid and (±)-linalool. *J Insect Physiol.* 1996;42:565–578.
30. Todd J, Baker T. Antennal lobe partitioning of behaviorally active odors in female cabbage looper moths. *Naturwissenschaften.* 1996;83:324–326.
31. Sadek MM, Hansson BS, Rospars JP, Anton S. Glomerular representation of plant volatiles and sex pheromone components in the antennal lobe of the female *Spodoptera littoralis. J Exp Biol.* 2002;205:1363–1376.
32. Hillier NK, Kleineidam C, Vickers NJ. Physiology and glomerular projections of olfactory receptor neurons on the antenna of female *Heliothis virescens* (Lepidoptera: Noctuidae) responsive to behaviorally relevant odors. *J Comp Physiol A.* 2006;192: 199–219.
33. Keil TA. Fine structure of the pheromone-sensitive sensilla on the antenna of the hawkmoth, *Manduca sexta. Tissue Cell.* 1989;21:139–151.
34. Steinbrecht R. Zur morphometrie der antenne des seidenspinners, *Bombyx mori* L.: Zahl und Verteilung der Riechsensillen (Insecta, Lepidoptera). *Z Morphol Tiere.* 1970;68:93–126.
35. Den Otter CJ, Schuil HA, Oosten AS-V. Reception of host-plant odours and female sex pheromone in *Adoxophyes orana* (Lepidoptera: Tortricidae): electrophysiology and morphology. *Entomol Exp Appl.* 1978;24:570–578.
36. Berg BG, Almaas TJ, Bjaalie JG, Mustaparta H. Projections of male-specific receptor neurons in the antennal lobe of the Oriental tobacco budworm moth, *Helicoverpa assulta*: a unique glomerular organization among related species. *J Comp Neurol.* 2005;486:209–220.
37. Zhao XC, Berg BG. Morphological and physiological characteristics of the serotonin-immunoreactive neuron in the antennal lobe of the male oriental tobacco budworm, *Helicoverpa assulta. Chem Senses.* 2009;34:363–372.
38. Cossé AA, Todd JL, Baker TC. Neurons discovered in male *Helicoverpa zea* antennae that correlate with pheromone-mediated attraction and interspecific antagonism. *J Comp Physiol A.* 1998;182:585–594.

39. Baker TC, Ochieng SA, Cosse AA, et al. A comparison of responses from olfactory receptor neurons of *Heliothis subflexa* and *Heliothis virescens* to components of their sex pheromone. *J Comp Physiol A Neuroethol Sens Neural Behav Physiol.* 2004;190: 155–165.
40. Kanaujia S, Kaissling KE. Interactions of pheromone with moth antennae: adsorption, desorption and transport. *J Insect Physiol.* 1985;31:71–81.
41. Zhou JJ. Odorant-binding proteins in insects. *Vitam Horm.* 2010;83:241–272.
42. Leal WS. Odorant reception in insects: roles of receptors, binding proteins, and degrading enzymes. *Annu Rev Entomol.* 2013;58:373–391.
43. Sakurai T, Namiki S, Kanzaki R. Molecular and neural mechanisms of sex pheromone reception and processing in the silkmoth. *Front Physiol.* 2014;5:125.
44. Gao Q, Chess A. Identification of candidate *Drosophila* olfactory receptors from genomic DNA sequence. *Genomics.* 1999;60:31–39.
45. Vosshall LB, Amrein H, Morozov PS, Rzhetsky A, Axel R. A spatial map of olfactory receptor expression in the *Drosophila* antenna. *Cell.* 1999;96:725–736.
46. Sakurai T, Nakagawa T, Mitsuno H, et al. Identification and functional characterization of a sex pheromone receptor in the silkmoth *Bombyx mori. Proc Natl Acad Sci USA.* 2004;101:16653–16658.
47. Nakagawa T, Sakurai T, Nishioka T, Touhara K. Insect sex-pheromone signals mediated by specific combinations of olfactory receptors. *Science.* 2005;307:1638–1642.
48. Krieger J, Grosse-Wilde E, Gohl T, et al. Genes encoding candidate pheromone receptors in a moth (*Heliothis virescens*). *Proc Natl Acad Sci USA.* 2004;101:11845–11850.
49. Grosse-Wilde E, Gohl T, Bouche E, Breer H, Krieger J. Candidate pheromone receptors provide the basis for the response of distinct antennal neurons to pheromonal compounds. *Eur J Neurosci.* 2007;25:2364–2373.
50. Mitsuno H, Sakurai T, Murai M, et al. Identification of receptors of main sex-pheromone components of three lepidopteran species. *Eur J Neurosci.* 2008;28: 893–902.
51. Forstner M, Breer H, Krieger J. A receptor and binding protein interplay in the detection of a distinct pheromone component in the silkmoth *Antheraea polyphemus. Int J Biol Sci.* 2009;5:745–757.
52. Miura N, Nakagawa T, Tatsuki S, Touhara K, Ishikawa Y. A male-specific odorant receptor conserved through the evolution of sex pheromones in *Ostrinia* moth species. *Int J Biol Sci.* 2009;5:319–330.
53. Miura N, Nakagawa T, Touhara K, Ishikawa Y. Broadly and narrowly tuned odorant receptors are involved in female sex pheromone reception in *Ostrinia* moths. *Insect Biochem Mol Biol.* 2010;40:64–73.
54. Wang G, Vasquez GM, Schal C, Zwiebel LJ, Gould F. Functional characterization of pheromone receptors in the tobacco budworm *Heliothis virescens. Insect Mol Biol.* 2011;20:125–133.
55. Wanner KW, Nichols AS, Allen JE, et al. Sex pheromone receptor specificity in the European corn borer moth, *Ostrinia nubilalis. PLoS One.* 2010;5:e8685.
56. Montagne N, Chertemps T, Brigaud I, et al. Functional characterization of a sex pheromone receptor in the pest moth *Spodoptera littoralis* by heterologous expression in *Drosophila. Eur J Neurosci.* 2012;36:2588–2596.
57. Liu C, Liu Y, Walker WB, Dong S, Wang G. Identification and functional characterization of sex pheromone receptors in beet armyworm *Spodoptera exigua* (Hubner). *Insect Biochem Mol Biol.* 2013;43:747–754.
58. Liu Y, Liu C, Lin K, Wang G. Functional specificity of sex pheromone receptors in the cotton bollworm *Helicoverpa armigera. PLoS One.* 2013;8:e62094.
59. Sun M, Liu Y, Walker WB, et al. Identification and characterization of pheromone receptors and interplay between receptors and pheromone binding proteins in the diamondback moth, *Plutella xyllostella. PLoS One.* 2013;8:e62098.

60. Zhang DD, Lofstedt C. Functional evolution of a multigene family: orthologous and paralogous pheromone receptor genes in the turnip moth, *Agrotis segetum*. *PLoS One*. 2013;8:e77345.
61. Jiang XJ, Guo H, Di C, et al. Sequence similarity and functional comparisons of pheromone receptor orthologs in two closely related *Helicoverpa* species. *Insect Biochem Mol Biol*. 2014;48:63–74.
62. Syed Z, Ishida Y, Taylor K, Kimbrell DA, Leal WS. Pheromone reception in fruit flies expressing a moth's odorant receptor. *Proc Natl Acad Sci USA*. 2006;103: 16538–16543.
63. Grosse-Wilde E, Svatos A, Krieger J. A pheromone-binding protein mediates the bombykol-induced activation of a pheromone receptor in vitro. *Chem Senses*. 2006;31:547–555.
64. Vetter RS, Baker TC. Behavioral responses of male *Heliothis virescens* in a sustained-flight tunnel to combinations of seven compounds identified from female sex pheromone glands. *J Chem Ecol*. 1983;9:747–759.
65. Groot AT, Estock ML, Horovitz JL, et al. QTL analysis of sex pheromone blend differences between two closely related moths: insights into divergence in biosynthetic pathways. *Insect Biochem Mol Biol*. 2009;39:568–577.
66. Dobritsa AA, van der Goes van Naters W, Warr CG, Steinbrecht RA, Carlson JR. Integrating the molecular and cellular basis of odor coding in the *Drosophila* antenna. *Neuron*. 2003;37:827–841.
67. Syed Z, Kopp A, Kimbrell DA, Leal WS. Bombykol receptors in the silkworm moth and the fruit fly. *Proc Natl Acad Sci USA*. 2010;107:9436–9439.
68. Krieger J, Grosse-Wilde E, Gohl T, Breer H. Candidate pheromone receptors of the silkmoth *Bombyx mori*. *Eur J Neurosci*. 2005;21:2167–2176.
69. Suckling DM, Karg G, Gibb AR, Bradley SJ. Electroantennogram and oviposition responses of *Epiphyas postvittana* (Lepidoptera: Tortricidae) to plant volatiles. *NZ J Crop Hort Sci*. 1996;24:323–333.
70. Jordan MD, Anderson A, Begum D, et al. Odorant receptors from the light brown apple moth (*Epiphyas postvittana*) recognize important volatile compounds produced by plants. *Chem Senses*. 2009;34:383–394.
71. Hao H, Wei J, Dai J, Du J. Host-seeking and blood-feeding behavior of Aedes albopictus (Diptera: Culicidae) exposed to vapors of geraniol, citral, citronellal, eugenol, or anisaldehyde. *J Med Entomol*. 2008;45(3):533–539.
72. Semmler M, Abdel-Ghaffar F, Schmidt J, Mehlhorn H. Evaluation of biological and chemical insect repellents and their potential adverse effects. *Parasitol Res*. 2014;113(1):185–188.
73. Bengtsson JM, Gonzalez F, Cattaneo AM, et al. A predicted sex pheromone receptor of codling moth *Cydia pomonella* detects the plant volatile pear ester. *Front Ecol Evol*. 2014;2:33.
74. Knight A, Light D. Use of ethyl and propyl (E, Z)-2, 4-decadienoates in codling moth management: improved monitoring in Bartlett pear with high dose lures. *J Entomol Soc B C*. 2004;101:45–52.
75. Light DM, Knight AL, Henrick CA, et al. A pear-derived kairomone with pheromonal potency that attracts male and female codling moth, *Cydia pomonella* (L.). *Naturwissenschaften*. 2001;88:333–338.
76. Vogt RG, Riddiford LM. Pheromone binding and inactivation by moth antennae. *Nature*. 1981;293:161–163.
77. Gong DP, Zhang HJ, Zhao P, Xia QY, Xiang ZH. The odorant binding protein gene family from the genome of silkworm, *Bombyx mori*. *BMC Genomics*. 2009;10:332.
78. Gu SH, Zhou JJ, Wang GR, Zhang YJ, Guo YY. Sex pheromone recognition and immunolocalization of three pheromone binding proteins in the black cutworm moth *Agrotis ipsilon*. *Insect Biochem Mol Biol*. 2013;43:237–251.

79. Krieger J, von Nickisch-Rosenegk E, Mameli M, Pelosi P, Breer H. Binding proteins from the antennae of *Bombyx mori*. *Insect Biochem Mol Biol*. 1996;26:297–307.
80. Sun M, Liu Y, Wang G. Expression patterns and binding properties of three pheromone binding proteins in the diamondback moth, *Plutella xyllotella*. *J Insect Physiol*. 2013;59:46–55.
81. Liu NY, Liu CC, Dong SL. Functional differentiation of pheromone-binding proteins in the common cutworm *Spodoptera litura*. *Comp Biochem Physiol A Mol Integr Physiol*. 2013;165:254–262.
82. Klein U. Sensillum-lymph proteins from antennal olfactory hairs of the moth *Antheraea polyphemus* (Saturniidae). *Insect Biochem*. 1987;17:1193–1204.
83. Pelosi P, Zhou JJ, Ban LP, Calvello M. Soluble proteins in insect chemical communication. *Cell Mol Life Sci*. 2006;63:1658–1676.
84. Forstner M, Gohl T, Breer H, Krieger J. Candidate pheromone binding proteins of the silkmoth *Bombyx mori*. *Invert Neurosci*. 2006;6:177–187.
85. Prestwich GD. Proteins that smell: pheromone recognition and signal transduction. *Bioorg Med Chem*. 1996;4:505–513.
86. Pophof B. Pheromone-binding proteins contribute to the activation of olfactory receptor neurons in the silkmoths *Antheraea polyphemus* and *Bombyx mori*. *Chem Senses*. 2004;29:117–125.
87. Leal WS. Rapid binding, release and inactivation of insect pheromones. *Comp Biochem Physiol A Mol Integr Physiol*. 2007;148:S81.
88. Kaissling KE. Olfactory perireceptor and receptor events in moths: a kinetic model revised. *J Comp Physiol A Neuroethol Sens Neural Behav Physiol*. 2009;195:895–922.
89. Laughlin JD, Ha TS, Jones DN, Smith DP. Activation of pheromone-sensitive neurons is mediated by conformational activation of pheromone-binding protein. *Cell*. 2008;133:1255–1265.
90. Lautenschlager C, Leal WS, Clardy J. *Bombyx mori* pheromone-binding protein binding nonpheromone ligands: implications for pheromone recognition. *Structure*. 2007;15: 1148–1154.
91. Steinbrecht RA. Odorant-binding proteins: expression and function. *Ann NY Acad Sci*. 1998;855:323–332.
92. Ziegelberger G. Redox-shift of the pheromone-binding protein in the silkmoth *Antheraea polyphemus*. *Eur J Biochem*. 1995;232:706–711.
93. Steinbrecht RA, Laue M, Ziegelberger G. Immunolocalization of pheromone-binding protein and general odorant-binding protein in olfactory sensilla of the silk moths *Antheraea* and *Bombyx*. *Cell Tissue Res*. 1995;282:203–217.
94. Wojtasek H, Leal WS. Conformational change in the pheromone-binding protein from *Bombyx mori* induced by pH and by interaction with membranes. *J Biol Chem*. 1999;274:30950–30956.
95. Horst R, Damberger F, Luginbuhl P, et al. NMR structure reveals intramolecular regulation mechanism for pheromone binding and release. *Proc Natl Acad Sci USA*. 2001;98:14374–14379.
96. Michel E, Damberger FF, Ishida Y, et al. Dynamic conformational equilibria in the physiological function of the *Bombyx mori* pheromone-binding protein. *J Mol Biol*. 2011;408:922–931.
97. Sandler BH, Nikonova L, Leal WS, Clardy J. Sexual attraction in the silkworm moth: structure of the pheromone-binding-protein-bombykol complex. *Chem Biol*. 2000;7: 143–151.
98. Xu W, Xu X, Leal WS, Ames JB. Extrusion of the C-terminal helix in navel orangeworm moth pheromone-binding protein (AtraPBP1) controls pheromone binding. *Biochem Biophys Res Commun*. 2011;404:335–338.

99. Xu X, Xu W, Rayo J, et al. NMR structure of navel orangeworm moth pheromone-binding protein (AtraPBP1): implications for pH-sensitive pheromone detection. *Biochemistry*. 2010;49:1469–1476.
100. Damberger FF, Ishida Y, Leal WS, Wuthrich K. Structural basis of ligand binding and release in insect pheromone-binding proteins: NMR structure of *Antheraea polyphemus* PBP1 at pH 4.5. *J Mol Biol*. 2007;373:811–819.
101. Wang G, Carey AF, Carlson JR, Zwiebel LJ. Molecular basis of odor coding in the malaria vector mosquito *Anopheles gambiae*. *Proc Natl Acad Sci USA*. 2010;107:4418–4423.
102. Xu P, Atkinson R, Jones DN, Smith DP. *Drosophila* OBP LUSH is required for activity of pheromone-sensitive neurons. *Neuron*. 2005;45:193–200.
103. Gomez-Diaz C, Reina JH, Cambillau C, Benton R. Ligands for pheromone-sensing neurons are not conformationally activated odorant binding proteins. *PLoS Biol*. 2013;11:e1001546.
104. Maida R, Ziegelberger G, Kaissling KE. Ligand binding to six recombinant pheromone-binding proteins of *Antheraea polyphemus* and *Antheraea pernyi*. *J Comp Physiol B*. 2003;173:565–573.
105. Mohl C, Breer H, Krieger J. Species-specific pheromonal compounds induce distinct conformational changes of pheromone binding protein subtypes from *Antheraea polyphemus*. *Invert Neurosci*. 2002;4:165–174.
106. Bette S, Breer H, Krieger J. Probing a pheromone binding protein of the silkmoth Antheraea polyphemus by endogenous tryptophan fluorescence. *Insect Biochem Mol Biol*. 2002;32:241–246.
107. Xu P, Hooper AM, Pickett JA, Leal WS. Specificity determinants of the silkworm moth sex pheromone. *PLoS One*. 2012;7:e44190.
108. Laue M, Steinbrecht RA, Ziegelberger G. Immunocytochemical localization of general odorant-binding protein in olfactory sensilla of the silkmoth *Antheraea polyphemus*. *Naturwissenschaften*. 1994;81:178–180.
109. Pelosi P, Maida R. Odorant-binding proteins in insects. *Comp Biochem Physiol B Biochem Mol Biol*. 1995;111:503–514.
110. Wang GR, Wu KM, Guo YY. Cloning, expression and immunocytochemical localization of a general odorant-binding protein gene from *Helicoverpa armigera* (Hubner). *Insect Biochem Mol Biol*. 2003;33:115–124.
111. Zhang S-g, Maida R, Steinbrecht RA. Immunolocalization of odorant-binding proteins in noctuid moths (Insecta, Lepidoptera). *Chem Senses*. 2001;26:885–896.
112. Jacquin-Joly E, Bohbot J, Francois MC, Cain AH, Nagnan-Le Meillour P. Characterization of the general odorant-binding protein 2 in the molecular coding of odorants in *Mamestra brassicae*. *Eur J Biochem*. 2000;267:6708–6714.
113. Gong ZJ, Zhou WW, Yu HZ, et al. Cloning, expression and functional analysis of a general odorant-binding protein 2 gene of the rice striped stem borer, *Chilo suppressalis* (Walker) (Lepidoptera: Pyralidae). *Insect Mol Biol*. 2009;18:405–417.
114. Zhou JJ, Robertson G, He X, et al. Characterisation of *Bombyx mori* odorant-binding proteins reveals that a general odorant-binding protein discriminates between sex pheromone components. *J Mol Biol*. 2009;389:529–545.
115. Liu Z, Vidal DM, Syed Z, Ishida Y, Leal WS. Pheromone binding to general odorant-binding proteins from the navel orangeworm. *J Chem Ecol*. 2010;36:787–794.
116. Yin J, Feng H, Sun H, et al. Functional analysis of general odorant binding protein 2 from the meadow moth, *Loxostege sticticalis* L. (Lepidoptera: Pyralidae). *PLoS One*. 2012;7:e33589.
117. Rogers ME, Sun M, Lerner MR, Vogt RG. Snmp-1, a novel membrane protein of olfactory neurons of the silk moth *Antheraea polyphemus* with homology to the CD36 family of membrane proteins. *J Biol Chem*. 1997;272:14792–14799.

118. Vogt RG. Biochemical diversity of odor detection: OBPs, ODEs and SNMPs. In: Blomquist G, Vogt RG, eds. *Insect Pheromone Biochemistry and Molecular Biology*. London: Elsevier Academic Press; 2003:391–445.
119. Benton R, Vannice KS, Vosshall LB. An essential role for a CD36-related receptor in pheromone detection in *Drosophila*. *Nature*. 2007;450:289–293.
120. Forstner M, Gohl T, Gondesen I, et al. Differential expression of SNMP-1 and SNMP-2 proteins in pheromone-sensitive hairs of moths. *Chem Senses*. 2008;33:291–299.
121. Rasmussen JT, Berglund L, Rasmussen MS, Petersen TE. Assignment of disulfide bridges in bovine CD36. *Eur J Biochem*. 1998;257:488–494.
122. Levy E, Spahis S, Sinnett D, et al. Intestinal cholesterol transport proteins: an update and beyond. *Curr Opin Lipidol*. 2007;18:310–318.
123. Nassir F, Wilson B, Han X, Gross RW, Abumrad NA. CD36 is important for fatty acid and cholesterol uptake by the proximal but not distal intestine. *J Biol Chem*. 2007;282: 19493–19501.
124. Febbraio M, Silverstein RL. CD36: implications in cardiovascular disease. *Int J Biochem Cell Biol*. 2007;39:2012–2030.
125. Robertson HM, Martos R, Sears CR, et al. Diversity of odorant binding proteins revealed by an expressed sequence tag project on male *Manduca sexta* moth antennae. *Insect Mol Biol*. 1999;8:501–518.
126. Rogers ME, Krieger J, Vogt RG. Antennal SNMPs (sensory neuron membrane proteins) of Lepidoptera define a unique family of invertebrate CD36-like proteins. *J Neurobiol*. 2001;49:47–61.
127. Nichols Z, Vogt RG. The SNMP/CD36 gene family in Diptera, Hymenoptera and Coleoptera: *Drosophila melanogaster, D. pseudoobscura, Anopheles gambiae, Aedes aegypti, Apis mellifera*, and *Tribolium castaneum*. *Insect Biochem Mol Biol*. 2008;38:398–415.
128. Vogt RG, Miller NE, Litvack R, et al. The insect SNMP gene family. *Insect Biochem Mol Biol*. 2009;39:448–456.
129. Li P-Y, Qin Y-C. Molecular cloning and characterization of sensory neuron membrane protein and expression pattern analysis in the diamondback moth, *Plutella xylostella* (Lepidoptera: Plutellidae). *Appl Entomol Zool*. 2011;46:497–504.
130. Gu SH, Yang RN, Guo MB, et al. Molecular identification and differential expression of sensory neuron membrane proteins in the antennae of the black cutworm moth *Agrotis ipsilon*. *J Insect Physiol*. 2013;59:430–443.
131. Liu S, Zhang YR, Zhou WW, et al. Identification and characterization of two sensory neuron membrane proteins from *Cnaphalocrocis medinalis* (Lepidoptera: Pyralidae). *Arch Insect Biochem Physiol*. 2013;82:29–42.
132. Liu C, Zhang J, Liu Y, Wang G, Dong S. Expression of SNMP1 and SNMP2 genes in antennal sensilla of *Spodoptera exigua* (Hubner). *Arch Insect Biochem Physiol*. 2014;85: 114–126.
133. Zhang J, Liu Y, Walker WB, Dong SL, Wang GR. Identification and localization of two sensory neuron membrane proteins from *Spodoptera litura* (Lepidoptera: Noctuidae). *Insect Sci*. 2014 (in press).
134. Krieger J, Raming K, Dewer YM, et al. A divergent gene family encoding candidate olfactory receptors of the moth *Heliothis virescens*. *Eur J Neurosci*. 2002;16:619–628.
135. Jin X, Ha TS, Smith DP. SNMP is a signaling component required for pheromone sensitivity in *Drosophila*. *Proc Natl Acad Sci USA*. 2008;105:10996–11001.
136. Fukuwatari T, Kawada T, Tsuruta M, et al. Expression of the putative membrane fatty acid transporter (FAT) in taste buds of the circumvallate papillae in rats. *FEBS Lett*. 1997;414:461–464.
137. Gaillard D, Passilly-Degrace P, Besnard P. Molecular mechanisms of fat preference and overeating. *Ann NY Acad Sci*. 2008;1141:163–175.

138. Boeckh J, Tolbert LP. Synaptic organization and development of the antennal lobe in insects. *Microsc Res Tech.* 1993;24:260–280.
139. Ernst KD, Boeckh J. A neuroanatomical study on the organization of the central antennal pathways in insects. III. Neuroanatomical characterization of physiologically defined response types of deutocerebral neurons in *Periplaneta americana. Cell Tissue Res.* 1983;229:1–22.
140. Vickers NJ, Christensen TA, Hildebrand JG. Combinatorial odor discrimination in the brain: attractive and antagonist odor blends are represented in distinct combinations of uniquely identifiable glomeruli. *J Comp Neurol.* 1998;400:35–56.
141. Berg BG, Galizia CG, Brandt R, Mustaparta H. Digital atlases of the antennal lobe in two species of tobacco budworm moths, the Oriental *Helicoverpa assulta* (male) and the American *Heliothis virescens* (male and female). *J Comp Neurol.* 2002;446:123–134.
142. Vickers NJ, Christensen TA. Functional divergence of spatially conserved olfactory glomeruli in two related moth species. *Chem Senses.* 2003;28:325–338.
143. Hansson BS, Ljungberg H, Hallberg E, Lofstedt C. Functional specialization of olfactory glomeruli in a moth. *Science.* 1992;256:1313–1315.
144. Hansson BS, Anton S. Function and morphology of the antennal lobe: new developments. *Annu Rev Entomol.* 2000;45:203–231.
145. Sakurai T, Mitsuno H, Haupt SS, et al. A single sex pheromone receptor determines chemical response specificity of sexual behavior in the silkmoth *Bombyx mori. PLoS Genet.* 2011;7:e1002115.
146. Tumlinson JH, Brennan MM, Doolittle RE, et al. Identification of a pheromone blend attractive to *Manduca sexta* (L.) males in a wind tunnel. *Arch Insect Biochem Physiol.* 1989;10:255–271.
147. Hansson BS, Christensen TA, Hildebrand JG. Functionally distinct subdivisions of the macroglomerular complex in the antennal lobe of the male sphinx moth *Manduca sexta. J Comp Neurol.* 1991;312:264–278.
148. Christensen T, Mustaparta H, Hildebrand J. Chemical communication in heliothine moths. *J Comp Physiol A.* 1991;169:259–274.
149. Hansson BS, Almaas TJ, Anton S. Chemical communication in heliothine moths. *J Comp Physiol A.* 1995;177:535–543.
150. Berg BG, Almaas TJ, Bjaalie JG, Mustaparta H. The macroglomerular complex of the antennal lobe in the tobacco budworm moth *Heliothis virescens*: specified subdivision in four compartments according to information about biologically significant compounds. *J Comp Physiol A Sens Neural Behav Physiol.* 1998;183:669–682.
151. Galizia CG, Sachse S, Mustaparta H. Calcium responses to pheromones and plant odours in the antennal lobe of the male and female moth *Heliothis virescens. J Comp Physiol A.* 2000;186:1049–1063.
152. Rospars JP. Invariance and sex-specific variations of the glomerular organization in the antennal lobes of a moth, *Mamestra brassicae*, and a butterfly, *Pieris brassicae. J Comp Neurol.* 1983;220:80–96.
153. Galizia CG, McIlwrath SL, Menzel R. A digital three-dimensional atlas of the honeybee antennal lobe based on optical sections acquired by confocal microscopy. *Cell Tissue Res.* 1999;295:383–394.
154. Rospars JP, Hildebrand JG. Sexually dimorphic and isomorphic glomeruli in the antennal lobes of the sphinx moth *Manduca sexta. Chem Senses.* 2000;25:119–129.
155. Skiri HT, Ro H, Berg BG, Mustaparta H. Consistent organization of glomeruli in the antennal lobes of related species of heliothine moths. *J Comp Neurol.* 2005;491:367–380.
156. de Bruyne M, Baker TC. Odor detection in insects: volatile codes. *J Chem Ecol.* 2008;34:882–897.

157. Couto A, Alenius M, Dickson BJ. Molecular, anatomical, and functional organization of the *Drosophila* olfactory system. *Curr Biol.* 2005;15:1535–1547.
158. Fishilevich E, Vosshall LB. Genetic and functional subdivision of the *Drosophila* antennal lobe. *Curr Biol.* 2005;15:1548–1553.
159. Grosse-Wilde E, Kuebler LS, Bucks S, et al. Antennal transcriptome of *Manduca sexta*. *Proc Natl Acad Sci USA.* 2011;108:7449–7454.
160. Daimon T, Fujii T, Fujii T, et al. Reinvestigation of the sex pheromone of the wild silkmoth *Bombyx mandarina*: the effects of bombykal and bombykyl acetate. *J Chem Ecol.* 2012;38:1031–1035.
161. Kehat M, Dunkelblum E. Behavioral responses of male *Heliothis armigera* (Lepidoptera: Noctuidae) moths in a flight tunnel to combinations of components identified from female sex pheromone glands. *J Insect Behav.* 1990;3:75–83.
162. Zhang JP, Salcedo C, Fang YL, Zhang RJ, Zhang ZN. An overlooked component: (Z)-9-tetradecenal as a sex pheromone in *Helicoverpa armigera*. *J Insect Physiol.* 2012;58:1209–1216.
163. Dunkelblum E, Gothilf S, Kehat M. Identification of the sex pheromone of the cotton bollworm, *Heliothis armigera*, in Israel. *Phytoparasitica.* 1980;8:209–211.
164. Witzgall P, Kirsch P, Cork A. Sex pheromones and their impact on pest management. *J Chem Ecol.* 2010;36:80–100.

INDEX

Note: Page numbers followed by "*f*" indicate figures and "*t*" indicate tables.

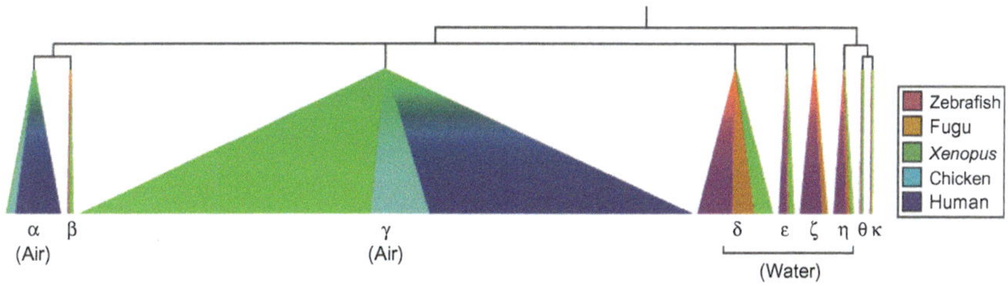

PLATE 1 ***Evolutionary dynamics of OR genes: a phylogenetic tree of OR genes from five vertebrate species*.** The genes that belong to different groups are represented by different colored triangles. The size of each triangle is approximately proportional to the number of OR genes from each species. The α and γ group genes are proposed to primarily detect airborne odorants because they exist in tetrapods, whereas the δ, ε, ζ, and η group genes that exist in fishes and Xenopus appear to primarily detect water-soluble odorants. The functions of the group β, θ, and κ genes are unclear. *Adapted by permission from Macmillan Publishers Ltd. Nature Reviews Genetics,*[29] *copyright 2008.* (Figure 1 on page 4 of this Volume)

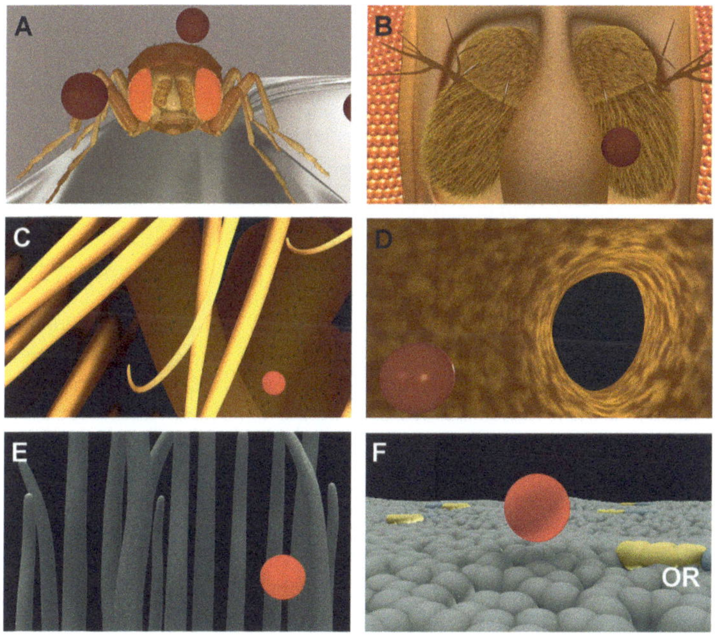

PLATE 2 ***Pathway of odor molecules travelling from the external environment to an olfactory receptor*.** Odor molecules symbolized by red spheres approach a fly (A) the paired antennae of which (B) are equipped with sensilla (C). The surface of these structures contains pores (C, D) that allow odor molecules to reach the interior of sensilla where the dendrites of olfactory sensory neurons are localized (E). The dendritic membrane contains heterodimeric receptors (ORs) that bind the odorant molecules (F). *Figures are video stills taken from http://vimeo.com/45493717 Copyright: © 2012 Kimberly Falk.* (Figure 1 on page 38 of this Volume)

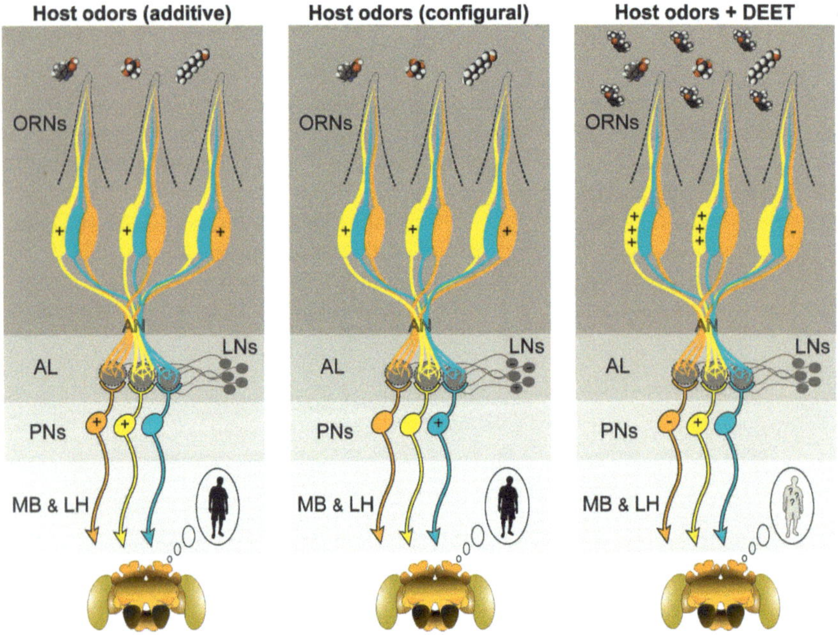

PLATE 3 ***Neural processing of olfactory cues in insect vectors.*** Odorants stimulate olfactory receptor neurons (ORNs) that have distinct phenotypes, as depicted by yellow, orange, and blue. Axons of ORNs of same phenotype coalesce into a single glomerulus in the antennal lobe (AL) region of the brain. Local interneurons (LNs) form synaptic connections between subsets of AL glomeruli, both stimulating and inhibiting AL responses. Projection neurons (PNs) transmit spatially segregated neural signals of the AL to higher brain centers including the mushroom bodies (MBs) and lateral horn (LH) for further processing. These circuits mediate behavioral responses to odors in insects. The three panels depict olfactory circuit logic possibilities in insects. (+) Indicates neural stimulation, (−) indicates neural inhibition, and (+++) indicates neural hyperstimulation. (Left panel) Additive processing in the AL shows the resulting signals from an odorous blend as the sum of the responses to individual components. (Center panel) Configural processing in the AL shows the resulting signals from an odorous blend as unique and not the sum of the responses to individual components. LNs may mediate AL processing. (Right panel) Repellents like DEET may alter ORN responses to host odors, potentially confusing the insect vector's ability to discriminate a host. These effects change the perception of insects, thus altering host-seeking behavior. (Figure 3 on page 91 of this Volume)

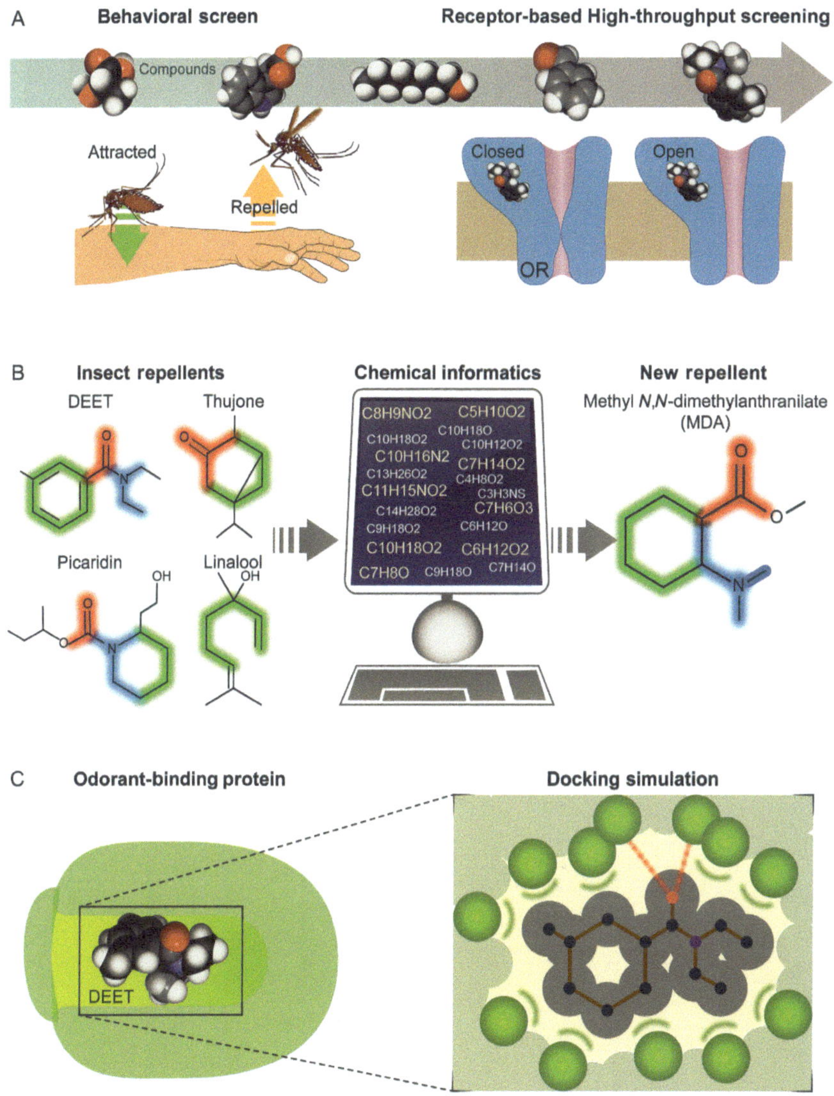

PLATE 4 ***Discovery pipeline of insect repellents.*** (A) The organism or a specific odorant receptor (OR) might be screened with thousands of chemical compounds to identify behavior- or receptor activity-altering chemicals. (B) Specific chemical descriptors associated with repellency from known insect repellents are established to predict future potential repellents.[181] (C) Amino acid residues interacting with the ligand (here DEET) are mapped using the known crystal structure of an odorant-binding protein. Hydrogen bonds (red dots) and hydrophobic interactions (green arcs) represent these interactions.[113] Mapping of the binding pocket may allow for the rational design of competitive ligand with behavior-modifying properties. (Figure 4 on page 97 of this Volume)

Papier | Fördert gute Waldnutzung
FSC® C083411

Kolibri 360 GmbH
Ferdinand-Jühlke-Straße 7
99095 Erfurt, Deutschland
produktsicherheit@kolibri360.de